Introduction to Oil and Gas Operational Safety

Introduction to Oil and Gas Operational Safety is aligned directly to t   ...ional Technical
Certificate in Oil and Gas Operational Safety.

Concisely written by a highly experienced team, this full-colour referen ᵣovides complete coverage of the
syllabus, including chapters on fire hazards, risk management and emergency response. It will ensure that you
are fully equipped with the knowledge and understanding to respond and deal with the daily hazards you may
face whilst working in the oil and gas industry.

Complete with tables, case studies and self-test questions, this book will guide you through the principles of
how to manage both offshore and onshore operational risks to prepare you for your exam and beyond.

Wise Global Training is a company dedicated to providing quality health and safety training in a variety of
formats, including eLearning and classroom-based courses as well as webinars.

Introduction to Oil and Gas Operational Safety

For the NEBOSH International Technical Certificate in Oil and Gas Operational Safety

Wise Global Training

Routledge
Taylor & Francis Group

LONDON AND NEW YORK

First edition published 2015
by Routledge
2 Park Square, Milton Park, Abingdon, Oxon OX14 4RN

and by Routledge
711 Third Avenue, New York, NY 10017

Routledge is an imprint of the Taylor & Francis Group, an informa business

British Library Cataloguing-in-Publication Data
A catalogue record for this book is available from the British Library

Library of Congress Cataloging-in-Publication Data
Introduction to oil and gas operational safety: for the NEBOSH international technical certificate in oil and gas operational safety / Wise Global Training.

pages cm

Includes bibliographical references and index.

1. Petroleum industry and trade–Safety measures–Examinations–Study guides.   2. Petroleum engineering–Safety measures–Examinations–Study guides.   3. Petroleum engineers–Certification.   4. National Examination Board in Occupational Safety and Health–Examinations–Study guides.

TP690.6I64 2014

338.2'728–dc23

2014014632

ISBN: 978-0-415-73077-8 (pbk)
ISBN: 978-1-315-84992-8 (ebk)

Typeset in Univers LT by
Servis Filmsetting Ltd, Stockport, Cheshire
Printed in Great Britain by Ashford Colour Press Ltd.

Contents

Contents

Preface

This book aims to provide a sound breadth of underpinned knowledge that will give the reader a foundation of understanding how to manage operational risks effectively within the oil and gas industry. It is aimed at those people who work, or who have aspirations to work, in the oil and gas sector either onshore or offshore.

This book mirrors the NEBOSH International Technical Certificate in Oil and Gas Operational Safety syllabus, and anyone wishing to study for this qualification will find this book a useful element of their learning programme. It is recommended that such a learning programme should also include a tutor-supported course to enhance understanding of the subject matter. This can be either an eLearning or classroom course where interaction with highly experienced tutors will potentially greatly improve examination pass rates. Wise Global Training, as the author of this book, provides both types of training, and our contact details can be found in the introduction to the Revision and Assessment Guide on page 189. The book's breadth of coverage and depth of understanding is the result of the quality of input from a team of highly qualified people with many years of hands-on experience working at the 'well head'. The book's style of writing portrays the author's desire to convey a technically challenging subject in a way that most people, regardless of background, will find comfortable to read and absorb. Extensive use of case studies and uncomplicated explanations of various issues reflect this style.

In order that the reader can assess their understanding of the issues covered, we have included a number of assessment questions at the end of each chapter, the indicative answers to which can be found in the Revision and Assessment Guide section.

Acknowledgements

Peter Donoghue, Managing Director of Wise Global Training Ltd. Peter has been a successful entrepreneur and business adviser for the past thirty-five years. He acknowledges that his greatest satisfaction has come from those occasions when he has been able to pass on to others knowledge and experience, knowing it will help them to improve their own situation or further their career. This was his main motivation for agreeing to manage the writing of this publication. Peter would like to take this opportunity to thank, on behalf of Wise Global Training Ltd, all those people mentioned here, without whose help this publication would never have seen the light of day.

Pamela Donoghue, Course Development Director of Wise Global Training Ltd. Pamela has contributed greatly in sourcing information, constructing and writing content, setting type standards and proofreading drafts at various stages of development.

Jennifer Newgass, Financial Director of Wise Global Training Ltd. Jennifer has contributed greatly in sourcing copyright permissions, constructing illustrations and assisting in resolving technical issues relating to images in general.

David Newgass, Operations Director of Wise Global Training Ltd. David has contributed greatly in providing the technical support necessary in the sourcing of information, the writing of the content and the communications needed to complete this publication. He has also assisted in sourcing and editing various images used in this publication.

Acknowledgements

Wise Global Training Ltd, as the author of this publication, would also like to acknowledge the contribution of the following people.

John W. Pond, CMIOSH MEI. We have been extremely fortunate to secure the services of John in the role of independent reviewer of all of the course content for this publication. He is a Chartered Health and Safety Professional, with particular experience of working with flammable and toxic materials in a multiple site environment. His career has included working for Esso Petroleum (Exxon Mobil), Kuwait Petroleum International (Q8) and the UK affiliate of Electricity de France (EDF Energy). John is currently working as an Independent Safety Consultant as well as being the current Principal Examiner for the NEBOSH International Oil and Gas Certificate. He was also one of the key developers of the Energy Institute Process Safety Management framework publication.

David Green, Managing Director of STS Business Services Ltd. David has provided significant amounts of support and guidance on oil and gas related content.

Antony Little, DipNEBOSH, GradIOSH, for providing technical advice and support.

Graham R. Fisher, SHE, NEBOSH, and **OPITO Trainer** for providing vast amounts of background information and technical support.

Kara Milne, Publisher with Routledge.

Joanna Endell-Cooper, Deputy Production Editorial Manager with Routledge.

Sadé Lee, Editorial Assistant with Routledge.

Thelma Gilbert, independent image researcher.

Illustrations credits

Illustrations credits

Tables

List of abbreviations

ABBI	Above, Below, Behind, Inside	ESDV	Emergency Shutdown Valve
ACGIH	American Conference of Government and Industrial Hygienists	EU	European Union
		F&G	Fire and Gas
ADR	European Agreement Concerning the International Carriage of Dangerous Goods by Road (Accord Européene Relatif au Transport International des Marchandises Dangereuses par Route)	FMEA	Failure Mode and Effect Analysis
		FMECA	Failure Modes and Effects Criticality Analysis
		FPS	Floating Production System
		FPSO	Floating Production Storage and Offloading unit
AFFF	Aqueous Film Forming Foam	FRC	Fast Rescue Craft
ALARP	As Low As Reasonably Practicable	GPA	General Platform Alarm
AMA	Advanced Medical Aid	GPS	Global Positioning System
Ar	Argon	GRP	Glass Reinforced Plastic
ATEX	Atmosphere Explosives. Also, the name commonly given to two European directives for controlling explosive atmospheres	GZ Drum	Gas Zone Drum
		H_2S	Hydrogen sulphide
		HAZID	Hazard Identification Study
BA	Breathing Apparatus	HAZOP	Hazard and Operability Study
BLEVE	Boiling Liquid Expanding Vapour Explosion	HDPE	High Density Polyethylene
B-O-P	Blow Out Preventer	HIPPS	High Integrity Pressure Protection System
BOSIET	Basic Offshore Safety Induction and Emergency Training	HLV	Heavy Lift Vessel
		HP	High Pressure
CCR	Central Control Room	HQ	Headquarters
CCTV	Closed Circuit Television	HSE	Health and Safety Executive
CFD	Computational Fluid Dynamics	HVAC	Heating, Ventilation and Air Conditioning
CNS	Central Nervous System	IDLH	Immediately Dangerous to Life or Health
CO_2	Carbon dioxide	IGC Code	International code for the construction and equipment of ships carrying liquefied gases in bulk
COMAH	Control of Major Accident Hazards		
COSC	Combined Operations Safety Case		
CRO	Control Room Operator		
CUI	Corrosion Under Insulation	IGG	Inert Gas Generator
CVCE	Confined Vapour Cloud Explosion	ILO	International Labour Organization
DCR	Design and Construction Regulations	IMO	International Maritime Organization
DNA	Deoxyribonucleic Acid	KW	Kilowatt
DP	Dynamic Positioning	kWm^2	Kilowatt per square metre
DPI	Dye Penetration Inspection	LEL	Lower Explosive Limit
DSEAR	Dangerous Substances and Explosive Atmospheres Regulations	LNG	Liquid Natural Gas
		LOTO	Lock Out Tag Out
DSV	Diving Support Vessel	LP	Low Pressure
DTLG	Decommissioning Technical Liaison Group	LPG	Liquid Petroleum Gas
EBS	Emergency Breathing Systems	LPI	Liquid Penetration Inspection
ECC	Emergency Command and Control	LSA	Low Specific Activity
EER	Evacuation, Escape and Rescue	MAPP	Major Accident Prevention Policy
ELD	Engineering Line Diagram	MCR	Maximum Continuous Rating
ELSA	Emergency Life Support Apparatus	MEDIVAC	Medical Evacuation
ERP	Emergency Response Procedure	MIG	Manual Inert Gas
ERRV	Emergency Response and Rescue Vessel	MOB	Man Overboard
ESD	Emergency Shutdown	MODU	Mobile Offshore Drilling Unit

Mpa	Mega Pascal	RT	Radiography Testing
MPGF	Multipoint Ground Flare	RTU	Remote Terminal Unit
MPI	Magnetic Particle Inspection	SALM	Single Anchor Leg Mooring
MSDS	Material Safety Data Sheet	SAR	Search and Rescue
N	Nitrogen	SBM	Single Buoy Mooring
NDT	Non-Destructive Testing	SBM	Synthetic Based Mud
NORM	Naturally Occurring Radioactive Materials	SBV	Standby Vessel
NSMC	North Sea Medical Centre	SCC	Stress Corrosion Cracking
NUI	Normally Unmanned Installation	SCE	Safety Critical Elements
OBM	Oil Based Mud	SIL	Safety Integrity Level
OIM	Offshore Installation Manager	SIMOPS	Simultaneous Operations
OSCR	Offshore Installations (Safety Case) Regulations 2005	SMART	Specific, Measurable, Achievable, Realistic and with Timescales
OSHA	Occupational Safety and Health Administration	SMS	Safety Management System
		SOLAS	Safety of Life at Sea
P&A	Plugging and Abandonment (of wells)	SOLAs	Specific Off-Label Approvals
P&ID	Piping and Instrument Diagram	SPM	Single Point Mooring
P&V	Pressure-Vacuum relief valve	SRB	Sulphate Reducing Bacteria
PAPA	Prepare to Abandon Platform Alarm	SSCV	Semi Submersible Crane Vessel
PFEER	Prevention of Fire, Explosion and Emergency Response	SSIV	Subsea Isolation Valve
		STEL	Short Term Exposure Limit
PFP	Passive Fire Protection	TDS	Total Dissolved Solids
pH	Power of hydrogen, i.e. a measure of how acidic a substance is	TEMPSC	Totally Enclosed Propelled Survival Craft
		TIG	Tungsten Inert Gas
PHA	Process Hazard Analysis	TLP	Tension Leg Platform
PIG	Pipeline Inspection Gauge	TLV	Time Limit Value
POB	Personnel On Board	TMT	Tube Metal Temperature
PPE	Personal Protective Equipment	TR	Temporary Refuge
ppm	Parts per million	Tremcard	Transport Emergency Card
PSIC	Protective Systems Isolation Certificate	TWA	Time Weighted Average
PSM	Process Safety Management	UEL	Upper Explosive Limit
PSV	Platform Support Vessel	UKOOA	United Kingdom Offshore Operators Association
PT	Penetrant Testing	UN	United Nations
PTW	Permit To Work	UNECE	United Nations Economic Commission for Europe
RBM	Risk Based Management		
RID	European Regulations Concerning the International Carriage of Dangerous Goods by Rail (Règlement Concernant le Transport International Ferroviare des Merchandises Dangereuses par Chemin de Fer)	UVCE	Unconfined Vapour Cloud Explosion
		VCF	Vapour Cloud Fire
		VDU	Visual Display Unit
		VESDA	Very Early Smoke Detection Apparatus
ROSOV	Remotely Operated Shut-Off Valve	VLSSCV	Very Large Semi-Submersible Crane Vessel
ROV	Remotely Operated Vehicle	VTC	Vocational Training Certificate
RPE	Respiratory Protective Equipment	WBM	Water Based Mud

Introduction to learning outcomes and command words

The syllabus set by NEBOSH for the International Technical Certificate in Oil and Gas Operational Safety is broken down into individual sections which have been translated into chapters within this publication. Each section, or chapter, has a clear learning outcome and command words are used in the learning outcome to indicate what is required of the student in relation to each item of content. For example, after completing Chapter 1 the reader should be able to 'Explain the purpose of, and procedures for, investigating incidents and how the lessons learnt can be used to improve health and safety in the oil and gas industries'.

Examination questions are set by NEBOSH to discover not only how much of a subject a student knows but also the associated skills that they are expected to demonstrate. Marks are then based on how effectively these skills are demonstrated. Command words are the guides in the question as to what assessment skill is being targeted by the question. These skills are generally knowledge, comprehension and application.

Knowledge requires an ability to recall or remember facts without necessarily understanding them. Command words used in knowledge based questions include **identify**.

Comprehension requires an ability to understand and interpret learned information. Command words used in comprehension based questions include **outline**.

Application is the skill of being able to take knowledge and apply it in different contexts and circumstances in order to understand why and where problems and issues arise. Command words used to assess the skill of application include **explain**.

The most frequently used command words include:

- **Identify** – This is asking for an answer which selects and names a subject. For example 'IDENTIFY three types of non-destructive testing of welds'.
- **Give** – This is asking for an answer without an explanation. For example 'GIVE an example of . . . '; 'GIVE the meaning of . . . '
- **Outline** – This is asking for an answer which supplies the principle features or different parts of a subject or issue. An exhaustive description is not required. What is sought is a brief summary of the major aspects of whatever is stated in the question.
- **Describe** – This is asking for a detailed written account of the distinctive features of a subject. The account should be factual without any attempt to explain. A further definition of DESCRIBE as a command word is 'a picture in word form'.
- **Explain** – This is asking for the reasoning behind, or an account of, a subject. The command word is testing the candidate's ability to know or understand why or how something happens.

How to use this book and what it covers

The *Introduction to Oil and Gas Operational Safety* is basically designed to:

1. cover the syllabus of the NEBOSH International Technical Certificate in Oil and Gas Operational Safety;
2. help students study, revise and sit the examination;
3. provide guidance for searching the internet and supply a range of significant websites;
4. provide a good updated reference text for supervisors, managers, safety representatives and newly qualified health and safety advisors within the oil and gas industries, both within and outside the UK.

We expect the book to be used as a basis for training and as further reference when students are back in their workplace. Although NEBOSH do not stipulate any prerequisites for undertaking this course, they do recommend that candidates have some form of health and safety qualification to act as a foundation of knowledge and understanding for the successful achievement of this qualification, such as the NEBOSH NGC or IGC.

Either of these courses can be studied with one of our sister publications: *Introduction to Health and Safety at Work NGC* or *Introduction to Health and Safety at Work IGC*, both written by Phil Hughes and Ed Ferrett.

The standard of English required by candidates studying for the NEBOSH International Technical Certificate in Oil and Gas Operational Safety must be such that they can both understand and articulate the concepts contained in the syllabus.

The syllabus for the NEBOSH International Technical Certificate in Oil and Gas Operational safety is divided into five elements, and how these relate to the publication is outlined in Table 0.1 below.

For more details see the NEBOSH syllabus guide at www.nebosh.org.uk

Table 0.1 Syllabus for the NEBOSH International Technical Certificate in Oil and Gas Operational Safety

Element no.	Chapters	Title	Recommended study hours
1	1–4	Health, safety and environmental management in context	12
2	5–10	Hydrocarbon process safety 1	8
3	11–16	Hydrocarbon process safety 2	8
4	17–18	Fire protection and emergency response	4
5	19–20	Logistics and transport operations	2
		Minimum total study time	34
		Recommended private study time	20
		Total overall hours	54

Sub-element 1.1:
learning from incidents

1.1 Investigating incidents, effective identification of the root causes and making recommendations for improvement

1.1.1 Why accidents and incidents should be investigated

There are two reasons why accidents and incidents should be investigated. The first reason is to determine their cause and the second is so that information forthcoming from the investigation can be used to prevent them happening again.

An investigation should not be used as a vehicle to apportion blame for an accident or incident. Employees should be encouraged to give their full co-operation to the investigating team, and this would perhaps be compromised if they thought that by doing so they might be held responsible in some way for the event.

There are risks and hazards within all workplaces, and measures are put in place to either eliminate them or reduce them to a level which is acceptable in order to prevent accidents or incidents which may be dangerous and/or result in injury, ill health or property damage. The very fact that an accident or incident has occurred means that those risk control measures current at the time are likely to have been less effective than expected. Furthermore, accidents and incidents need to be investigated for both legal and financial reasons.

1.1.2 Legal reasons for investigating accidents and incidents

▷ To demonstrate that the company is meeting its legal requirement to investigate accidents and incidents.

▷ Persons who have been affected by an accident may consider taking legal action against the company as a result. Consequently, employers should make available information regarding the circumstances appertaining to the accident which result from any investigation.

▷ If needs be, a company can demonstrate to the courts their commitment and positive attitude to health and safety by providing evidence of a thorough investigation of an accident which subsequently allowed them to take steps to put in place measures to prevent any future recurrence.

1.1.3 Financial reasons for investigating accidents and incidents

▷ Information forthcoming from an accident investigation provided to an insurance company may well assist in the event of a claim.

▷ Investigating dangerous incidents and rectifying the cause, such as an escape of flammable vapour, could prevent a recurrence with potential catastrophic results, thus averting the cost of repairs and replacement and lost revenue, as well as potentially saving lives and preventing injuries.

1.1.4 Other reasons for investigating accidents and incidents

The information and insights gained from the investigation of accidents and incidents are invaluable. The following are examples of the kinds of things to be expected when a thorough investigation is conducted:

▷ How and why an accident/incident happened.

▷ What working practices and procedures actually are – these may differ from what they should be (e.g. employees may take unacceptable risks to make their work easier or faster).

▷ How exposure to conditions (e.g. noise, cold, heat) or substances (e.g. chemicals, radiation, gases) may affect the health of employees.

▷ Weaknesses or faults in production systems whereby a certain scenario of events will expose these weaknesses or faults. For example, opening a flow valve too quickly might cause pressure to build up beyond a safe working pressure in a pipeline, resulting in an escape of flammable product from a flange. Remedial action would focus on safe working procedures, possible pressure relief systems or the strengthening of flanges.

▷ An investigation can highlight weaknesses in existing risk control measures and allow employers to review and supplement them in order to prevent accidents/incidents in the future. Lessons learned in one department of an organization can be used in other departments, thus benefitting the organization as a whole.

1.1.5 Benefits from investigating accidents and incidents

As well as information and insights being gained from investigating accidents and incidents, there are also tangible benefits to be had. The following are examples of such benefits:

▷ The outcomes of an investigation can result in the company putting measures in place to prevent the recurrence of similar accidents or incidents in the future.

▷ The company can avoid business losses if they take heed of the outcome of an investigation by preventing further accidents or incidents. For example, the escape of flammable product we cited earlier which, in this case, was caught before it escalated into a potentially catastrophic incident. The benefits of remedial action can be measured in the number of potential lives saved as well as the potential savings of the cost of repairs and replacement and lost revenue. Other costs saved might include the cost of legal action which might be taken against the company; increased insurance premiums; and loss of business due to a bad reputation, resulting in lost orders.

▷ The development of a health and safety culture within the company. Following an investigation, any measures which are put in place as a result of the findings will be more readily accepted by the workforce, especially if they were involved in the decision-making process.

▷ Managerial skills will be developed during any investigation and these can be used in other departments within the organization.

1.1.6 Near misses and dangerous occurrences

Not only is the investigation of accidents and incidents important, it is also useful to investigate near misses and dangerous occurrences which did not result in injury. Just because no-one has been injured on one occasion does not mean that if the event happened again the result would be the same.

Whether the incident is classed as an accident, a near miss or a dangerous occurrence, the investigation should carry the same degree of importance, and the findings will be as useful in any event in preventing a recurrence. Specific lessons should be noted to identify why control measures already in place failed to prevent the incident and what further measures should be introduced to rectify the situation. General lessons learned from one incident will also be useful throughout an organization to increase awareness about health and safety issues.

Let's take a look at case study 1.1 involving a dangerous occurrence.

1.1.7 Categorizing incidents

Accidents and incidents result in outcomes. If accidents/incidents are categorized under their various outcomes, it is possible to give them the appropriate degree of attention and/or investigation. For example, an accident where someone was hurt by tripping over a discarded box would not warrant the same level of investigation as an escape of hydrocarbon gas.

Let's first deal with the word **outcome**, to make it quite clear what we are talking about.

An **outcome** is the effect of an unplanned, uncontrolled event and can range from a mere escape of vapour which causes no ill effects and disperses into the atmosphere, to a major incident with structural damage and a number of injuries.

An outcome can result in one of the following:

▷ Injury
▷ Minor injury
▷ Significant injury
▷ Major injury
▷ Dangerous Occurrence or Damage Only
▷ Near miss

Let's look at what each of these outcomes means.

Outcome – injury

Injury is physical damage or harm which is suffered by a person.

▷ **Minor injury** – an injury which does not involve time off work.

Case Study 1.1

On a production platform, the process system was being started up and the start-up procedure was being followed. Suddenly, and without warning, a fire was detected around a pipe flange within an unmanned and confined section. The fire and gas detection system immediately detected the fire and emergency procedures were immediately activated. These included the Emergency Shutdown (ESD) system which isolates the affected section of the process so as to stop the fire being fed by further flammable product. The Emergency Shutdown (ESD) system also brings to an end the process system. The fire extinguisher system, which in this case uses an inert gas, was activated in that confined area to extinguish the fire.

The investigation that followed found a number of failings. These included the fact that the start-up procedure had not been followed precisely. One of the flow valves had been opened too quickly, allowing product to flow too quickly. This had two effects. First, pressure in the pipeline built up to a level which

came near to the maximum operating pressure for that pipeline. This was exacerbated by the fact that the flange connection bolts had not been checked during routine maintenance work and did not have the required torque tightness to hold the maximum pressure of the pipeline, resulting in the leak of flammable product.

Second, the increased flow of product also served to create a build-up of static charge in the pipeline. The electrical bonding of the pipeline was found to be insufficient and a spark was generated which ignited the leak from the flange, resulting in the fire.

Conclusions from the investigations were: (1) personnel involved in process systems, and in particular, start-up and shutdown, should be given more regular training, including a wider understanding of the consequences of certain actions; and (2) a review of the extent and frequency of the maintenance programme should be conducted, with a view to making it more effective.

▶ **Significant injury** – a significant injury is one which is not 'major' but which results in the injured person being away from work OR being unable to carry out their full range of normal duties.
▶ **Major injury** – major injuries are those which can be regarded as a serious threat to a person's health and/or well-being.

Outcome – Dangerous Occurrence or Damage Only

A Dangerous Occurrence or Damage Only is an event that happens which has not involved injury to anyone, but which could have done. The following list of examples is not meant to be exhaustive:

▶ A fire or explosion not resulting in injury
▶ An escape of flammable or toxic gas, vapour or fluid
▶ The collapse, overturning or failure of load-bearing parts of lifts and lifting equipment
▶ Explosion, collapse or bursting of any closed vessel or associated pipework
▶ Electrical short circuit or overload causing a discharge spark
▶ Failure of industrial radiography or irradiation equipment to de-energize or return to its safe position after the intended exposure period

Outcome – near miss

A near miss is any unplanned incident, accident or emergency which did not result in an injury. An example would be materials falling from scaffolding and almost hitting an employee underneath. A near miss must be reported to determine the cause in order for changes to be made to prevent it happening again.

Figure 1.1 Near miss sign
Source: Dreamstime.

1.1.8 Training for the investigating team

Any accident/incident investigation will involve a team of people with different skills depending on where they work. All will have had previous training in accident investigation. A typical team will include:

▶ Operations team leader
▶ Field or platform safety officer if it's an offshore installation
▶ Safety representatives
▶ Area authorities (responsible for the area where the incident took place)
▶ Senior onshore managers
▶ Specialist inspectors
▶ Tool pusher if it's a drilling rig

The investigating team will need to have information and training regarding:

▶ Their roles and responsibilities
▶ How to identify which events need to be reported
▶ How to complete documentation
▶ Accident book regulations and requirements and how to use it as a source of historical information
▶ Documents and forms relevant to the investigation – internal and external
▶ The importance of reporting accidents/incidents/ dangerous occurrences/near misses for legal, investigative and monitoring reasons
▶ The dissemination of information and to whom

1.1.9 The accident/incident investigation

There is a generally accepted four-stage process to investigating accidents/incidents.

▶ **Step one** – gathering the information
▶ **Step two** – analysing the information
▶ **Step three** – identifying the required risk control measures
▶ **Step four** – formulation of the action plan and its implementation

Let's look at each of these steps in more detail.

The accident/incident investigation step 1 – gathering the information

Some of the information required to conduct the investigation will be taken from the scene of the incident, which must be kept secure and undisturbed until the team is satisfied that they have all the facts they need. This could be when all photographs, sketches and measurements have been made.

If the incident is serious enough to involve the enforcing authorities and/or police, then the area must be secured until permission has been given by those authorities to open the site.

Should the area pose a risk to others, then the area must be made safe but photographs should be taken if possible before the site is disturbed.

The process of gathering the information should ensure that it:

▷ Explores all reasonable lines of enquiry
▷ Is timely – should be done as soon as possible after the event
▷ Is structured, setting out clearly what is known, what is not known and records the investigative process

This is the stage to collect information about:

▷ Where it happened
▷ When it happened
▷ Who was involved
▷ What was involved

Relevant information will include sketches, measurements, instrument readings and records, logged information, photographs, check sheets, permits-to-work and details of the environmental conditions at the time if relevant. Opinions, experiences and observations as well as events which led up to the accident/incident should also be recorded.

Gathering the information – observational techniques

Observation is an integral part of an investigation, and managers and supervisors will need to be trained in observational techniques. The observation of working procedures will identify those practices which are regarded as unsafe and which need to be changed in order to make them safe.

Good observation skills should include:

▷ Knowledge of the workplace and procedures
▷ Being open minded
▷ Keeping a systematic record of observations

Each of these skills can help to ensure observations are done effectively.

Good observation relies on a number of additional techniques. The observer should:

▷ Take time to observe the whole scene
▷ Be alert to possible changes to the accident scene by those who may have a motive to correct unsafe practices
▷ Use the ABBI technique: look Above, Below, Behind, and Inside
▷ Be inquisitive and question employees to determine risks – their views can be a valuable source of insight
▷ Use all senses including smell, sight, touch and hearing
▷ Have an open mind and look for solutions
▷ Identify, record and feedback good performance as well as bad

Gathering the information – interview techniques should include:

▷ Using an interviewing style which does not reflect a blame culture
▷ Asking questions in a way which does not make the interviewee feel intimidated or uncomfortable
▷ Conducting the interview in familiar surroundings as this will be less intimidating
▷ Encouraging co-operation by allowing witnesses to speak openly in their own words without using technical jargon
▷ Promoting a positive attitude to finding the reasons for the incident to prevent a recurrence in the future rather than apportioning blame for the present one
▷ Interviewing witnesses separately and in private to prevent them from influencing each other's accounts
▷ Providing a summary of what the witness said in order that they can ensure that everything has been understood correctly and that the interviewer has not misinterpreted the account

Gathering the information – plans

After the incident, any plans or sketches of where people were at the time as well as the layout of the area (e.g. machinery) will be helpful in finding the causes of the accident/incident.

Figure 1.2 Studying plans
Source: iStock.

Gathering the information – photographs

Photographs can provide evidence of the scenes of an incident immediately after the event. This is important because investigations can be ongoing and evidence may be lost.

Gathering the information – relevant records and sources of information

The amount of information and evidence required after an incident depends upon the seriousness of the outcome. They may include the following:

- Victim statements
- Witness statements
- Plans and diagrams
- CCTV coverage
- Process drawings, sketches, measurements, photographs
- Check sheets, permits-to-work records, method statements
- Details of the environmental conditions at the time
- Written instructions, procedures and risk assessments which should have been in operation and followed
- Previous accident records
- Information from health and safety meetings
- Technical information/guidance/toolbox talk sheets
- Manufacturers' instructions
- Risk assessments
- Training records
- Logs
- Instrument readouts and records
- Opinions, experiences, observations

The accident/incident investigation step 2 – analysing the information

Any analysis of the information collected should:

- Be objective and unbiased
- Identify the sequence of events and conditions that led up to the event
- Identify the immediate causes
- Identify underlying causes, i.e. actions in the past that have allowed or caused undetected unsafe conditions/practices
- Identify root causes, i.e. organizational and management health and safety arrangements – supervision, monitoring, training, resources allocated to health and safety, etc.

Analysis of this information should result in supplying reasons why the incident happened. It may also become clear what further information is still required. It is useful to analyse information gathered as an ongoing process during the investigation, as this will allow for other lines of enquiry to evolve and be developed.

By involving all members of the investigative team, differing opinions can be considered and a broad view of the results gained. The analysis should be conducted systematically to make sure nothing has been missed and that an unbiased outcome can be reached as to the immediate, underlying and root causes of the accident.

We shall be looking at these types of causes in more detail later in this chapter.

The accident/incident investigation step 3 – identifying the required risk control measures

Once the findings of the investigation have evolved, they will highlight failings in the existing control measures which led to the incident. They may also determine which control measures should be implemented in order to prevent a future recurrence.

Another outcome of an accident investigation is that it will prioritize the risk control measures to be implemented.

Usually those control measures which eliminate the risk by using engineering controls are more reliable than those controls which are reliant on people.

There is a 'hierarchy of risk control' which should be put in place in the following order:

Hierarchy of risk control

- Eliminate the risk altogether.
- Substitute the risk for something safer.
- Apply engineering controls such as cut out devices, guards, etc.
- Apply administrative controls such as safe working practices.
- Use Personal Protective Equipment (PPE), but only as a last resort or in conjunction with other controls.

The accident/incident investigation step 4 – formulation of the action plan and its implementation

The involvement of senior management in the formulation of an action plan will be necessary, as this level of management is generally the decision-making level within an organization. The investigation team will recommend additional risk control measures which have been determined as a result of the investigation. The action plan will determine which control measures should be implemented in the short term and others which will be long term measures.

The action plan should have **SMART** objectives, i.e. be:

- **S**pecific
- **M**easurable
- **A**chievable
- **R**ealistic, and with
- **T**imescales

The accident/incident investigation – analysing the causes

We will now take a more detailed look at the process of analysing the cause of an incident. Causes of incidents can be categorized into three basic types – immediate causes, underlying causes and root causes.

- **Immediate causes** are generally unsafe acts and/or conditions.

- **Underlying causes** are generally procedural failures.
- **Root causes** are generally management system failures.

Immediate cause

The most obvious reason why an adverse event happens, e.g. the guard is missing; the employee slips; the pipe flange fails,etc. There may be several immediate causes identified in any one adverse event.

Immediate causes are those which are responsible for the accident and are often easy to recognize.

Underlying cause

The less obvious 'system' or 'organizational' reason for an adverse event happening, e.g. pre-start-up machinery checks are not carried out; safe working procedures are not adhered to; the hazard has not been adequately considered via a suitable and sufficient risk assessment; production process pressures are too great, etc.

Root cause

Root causes are generally management, planning or organizational failings. For example, we mentioned safe working procedures not being adhered to, as an underlying cause. This could be because of lack of training or poor supervision, both of which are classed as root causes. Other examples of root causes include:

- A lack of rules and/or working procedures
- Insufficient training
- A general lack of commitment to safety
- Insufficient supervision
- Poor plant, equipment and layout design
- Poor working conditions

1.1.10 **Analysing causes of an accident**

There are many techniques used in determining the true picture of immediate and root causes of an accident/incident, one of which is by applying a questioning technique which constantly asks 'why?', and then applying the answers to a pictorial diagram which gives a causal tree analysis. If we apply this technique to the accident outlined below we can see how it works.

We start by looking at the ultimate consequence and work our way back by asking 'why?'. We can then get a full picture of all the causes.

Let's ask the questions 'why?' to our situation and see what transpires.

Joe is one of the control room operators on board a floating production storage and offloading (FPSO) platform.

He had been asked to inspect a fire and gas sensor some distance away from his normal work post. He decided to take a shortcut through an area that housed steam pipes.

Figure 1.3 Joe, accident victim
Source: iStock.

Unauthorized employees are not normally allowed in this area as the steam in the pipes is under pressure and of high temperature. As he was passing by the piping, one of the flanges of the steam pipes emitted a blast of steam which scalded Joe's hand and he suffered serious burns.

Figure 1.4 Scene of accident
Source: Dreamstime.

Joe was a walking wounded casualty and took himself to the medical bay for treatment to burns. An investigation into the incident was started.

Let's now work our way through the investigation of the accident involving Joe by asking a series of questions. These questions are meant to unfold the facts as they reveal themselves, much like the layers of an onion.

We start with the known fact that **'a person has been injured in an area housing steam pipes'**. From there we keep asking the question 'why?' in order to reveal the underlying cause(s) of each element as it is revealed. Let's have a look at how it works.

Fact – A person (Joe) has been injured in the area housing steam pipes.

Ask the question:
What was the cause?

Answer:
As Joe was passing one of the steam pipe flanges, there was an unpredicted blast of steam emitted from one of the flanges.

First of all we will look at the reason for Joe being in this area. We will look at the reasons for the emission of steam later.

Ask the question:
Why was Joe near the steam pipes?

Answer:
Because he took a shortcut.

Ask the question:
Was he allowed in that area?

Answer:
No. It was a restricted area.

Ask the question:
Why did he take an unauthorized shortcut?

Answer:
Because no-one had told him he could not.

Ask the question:
Why had no-one told him he could not take a shortcut through the area housing steam pipes?

Answer:
Lack of proper induction training and poor supervision.

Now we have to look at the reasons for the escape of steam.

Ask the question:
Why did the steam escape?

Answer:
Because there was a leak on the flange.

Ask the question:
Why was there a leak on the flange?

Answer:
Because some of the flange bolts had loosened.

Ask the question:
Why had the flange bolts loosened?

Answer:
Because the maintenance schedule had been allowed to lapse and the bolts had not been inspected in accordance with the schedule.

From this investigation, we can now build up a causal tree showing:

▶ **Immediate causes**
▶ **Underlying causes**
▶ **Root causes**

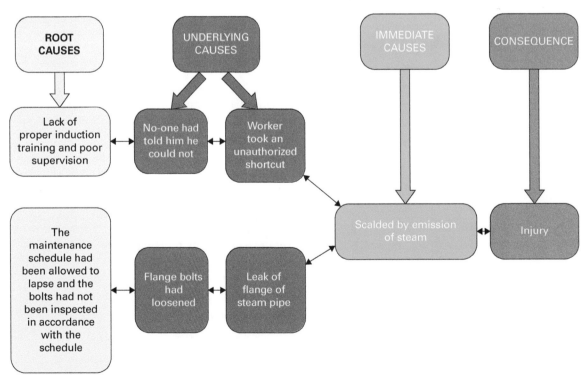

Figure 1.5 Causal tree diagram
Source: Wise Global Training.

As you can see, the causal tree we have been able to draw from constantly asking the question 'why?' has given us a clear picture of what all the true causes of the accident were.

This leaves us with a clear indication of the actions which need to be taken if we want to ensure there is no repetition of any of the events leading up to the accident.

1.1.11 Accident/incident report forms

There are many kinds of accident/incident report forms but all do the same job – they all include the findings of the investigation and determine the causes of the incident. They also provide recommendations to prevent further occurrences. There are also various computer programs which have been developed to record and analyse data. Whatever the format, they all state:

- **What happened –** the injuries/losses/costs
- **How it happened –** the event itself
- **Why it happened –** the causes: root, underlying and immediate
- **Recommendations –** any action to be taken to remedy the situation and prevent any recurrences

The use of standardized report forms ensures that the investigation process is correctly adhered to and that information can be reported back to management. Follow-up actions can easily be taken following appropriate recommendations within the report. Standardized report forms can also act as a checklist.

An efficient recording system will:

- Ensure the information is correctly and accurately presented
- Allow the data to be analysed easily in order to discover common causes or trends
- Ensure data which may be required for future reference is included
- Identify issues which may help prevent any recurrence of the accident

Report forms should be reviewed on a regular basis to ensure that any recommendations have been implemented.

1.2 The importance of learning lessons from major incidents, management, cultural and technical failures that may lead to such incidents

1.2.1 Learning the lessons

The lessons learned from accidents and incidents all contribute to the building up of knowledge and experience which can help in avoiding a repetition of such events. This is because the consequences of accidents in the oil and gas industry can be catastrophic, e.g. Deepwater Horizon, Piper Alpha, Buncefield, etc.

Acquiring knowledge and experience from accidents and incidents should be a structured process, whether it is a minor accident or a major incident. It is important that lessons learned stem from a basic understanding of the incident in order to develop a safety culture which can be vital in avoiding a major catastrophe.

Learning lessons from an incident can benefit two main areas:

1 Locally – the people directly involved, in order to prevent a recurrence of the incident
2 More widely – to ensure that the experience and lessons learned from one incident involving one system can be shared with other systems, thus avoiding similar incidents in other areas

1.2.2 Learning lessons locally

At the end of the investigation there needs to be a dissemination of the conclusions arrived at from all of the information which has been collected and analysed. It is important that these are communicated in a way that is understood by everyone, in a format which is appropriate to the level necessary. For example:

- For the management of the organization – they will need a report which details:
  - what went wrong
  - the systems and procedure failures which were involved
  - how serious the incident could have been
  - how to avoid future incidents.
- For regulatory bodies, safety records, etc. – technical details of the investigation as well as its findings.
- For operators of the systems and procedures involved in the incident – the findings and recommendations of the investigation should be in a format which is understandable at their level. This will allow changes to be made in order to prevent future incidents. The format could be in the form of checklists, written procedures, etc.
- For incident investigators generally. This group of people have an ongoing need to broaden their knowledge and understanding of how things go wrong and how to encompass this knowledge in future investigations.

Training and educational materials should also be in an appropriate format tailored to the target audience. These could be in:

- A written format – guidance notes, checklists, written procedures. If checklists are used it is important that they are not used in isolation but alongside other materials which give explanations of each item listed in order to make full use of the background knowledge and experience gained from an investigation.

- Audio-visual aids to convey the information – these can be training videos, which for some people are preferable to other methods of teaching.
- Verbal presentations – face-to-face teaching sessions which include explanation of the investigation and its findings from both the perspective of the investigator(s) and those who were involved in the accident.
- Training exercises – these could be simulations of incidents which would include exercises on how to deal with actual events should they occur, or even scenarios based on previous incidents using the personal accounts of those involved.

To promote a culture of learning within an organization, it is important that any lessons learned, and experience gained, are used effectively when new projects are planned or existing ones modified. The experience of others can be used in such techniques as 'peer assist'. Peer assist is a technique whereby a group of people who have been involved in investigating incidents can share their experiences and lessons learned with others who may have to conduct investigations in the future.

1.2.3 Learning lessons more widely

Lessons learned within one organization can be disseminated widely throughout other organizations by the publication of information in trade or specialized journals or publications, or through internet websites.

People who prepare information to be disseminated need to make it interesting and well presented. The author should be able to:

- Highlight relevant points which are of interest to the target audience
- If using first-hand accounts, be concise so that the information has maximum impact
- Use examples of different incidents to illustrate similarities and differences, but avoid repetition in order to maintain interest in the article
- At the end of the article, summarize the information and provide a conclusion
- Supplement the article with trend information

People who have first-hand knowledge and experience of an incident and dealing with the aftermath can be instrumental in teaching others. They can be involved in running workshops and providing presentations or training sessions.

As well as individuals disseminating knowledge and experience internally within their companies, there should be systems in place whereby these people can share their experience and knowledge with other organizations and agencies in order to broaden the range of beneficiaries. This could take the form of trade activities or training sessions where representatives of companies attend and where incidents are discussed. The information can then be reported back to be shared within their own companies.

The contra position of learning lessons is where lessons are simply ignored or abandoned, and in these cases the consequences can be catastrophic.

In the Piper Alpha case, the positioning of dangerous operations away from personnel areas had been recognized as a move towards a safer operating environment. However, when gas production began on Piper Alpha, a propane condensate pump was positioned near the control room for pure convenience. This disregard for lessons learned proved to be a fatal error in the consequences that followed.

It is a part of safety requirements in the oil and gas industry that companies take into consideration safety information, not only from within their own organization, but from other organizations, when assessing risks.

REVISION QUESTIONS FOR ELEMENT 1

1 **Explain** what the main purposes of an accident/incident investigation are.
2 (a) **Identify** the **FOUR** steps in an accident/incident investigation.
   (b) When gathering information following an accident/incident, **identify SIX** sources of information.
3 (a) **Explain** what a 'near miss' is.
   (b) **Explain** why near misses should be investigated.

You can compare your answers with our guidance at the end of the book under the section entitled 'Revision and assessment guide'.

CHAPTER 2

Element 1

Sub-element 1.2:
Hazards inherent in oil and gas

This chapter covers the following learning outcome

Explain the hazards inherent in oil and gas arising from the extraction, storage and processing of raw materials and products.

2.1 Meaning and relevance of various words and phrases associated with hazards inherent in oil and gas

There are many hazards inherent in the oil and gas industry, and risk assessments are conducted to ensure that risk control measures are put in place to prevent the realization of those hazards and to maintain a safe working environment.

Consequently, it is important to know the meaning and relevance of the common terms used throughout the industry in relation to hazards in the oil and gas industry, and the following information covers some of the basic terms used.

2.1.1 Flash point

The **flash point** of a volatile liquid is the lowest temperature at which it can vaporize to form an ignitable mixture when mixed with air. Consequently, storing a fluid at a temperature below its flash point is an effective way of preventing ignitable vapours from forming.

Figure 2.1 Fire ball
Source: iStock.

The characteristics of a vapour are described using two terms. These are:

▶ **Vapour density**
▶ **Vapour pressure**

Let's look at these two terms more closely.

2.1.2 Vapour density

Vapour density is the measurement of how dense a vapour is in comparison with air.

Comparing a vapour's density with air indicates whether the vapour will rise or fall if it is released into the atmosphere. So, if we say air has a density of 1, then any vapour with a density below 1 will rise as it is lighter than air, and any vapour with a density above 1 will fall as it is heavier than air.

Propane, which is a type of Liquefied Petroleum Gas (LPG), has a density of 2.0, which makes it heavier than air. Consequently, when mixed with air, propane vapour will fall.

Whereas methane (liquefied natural gas) has a density of 0.717, which makes it lighter than air, so when mixed with air methane vapour will rise.

Consideration of vapour density is a vital factor in deciding where to position gas detection equipment, general ventilation requirements, etc.

2.1.3 Vapour pressure

The process of evaporation involves the molecules on the surface of a liquid. When the energy within these molecules is sufficient for those molecules to escape, they do so in the form of a vapour. This is known as **vapour pressure**.

Vapour pressure is measured in the standard units of pressure known as pascal (Pa). 1 Pascal is 1 newton per square metre.

The greater the vapour pressure, the faster this process takes place, which results in a greater concentration of vapour. A substance with a high vapour pressure at normal temperatures is often referred to as volatile.

2.1.4 Flammability

Vapour which is flammable presents the risk of an explosion. However, some vapours are more flammable than others and, as such, are categorized to indicate the level of risk involved.

The degree of flammability can be expressed as follows:

▶ Flammable

▶ Highly flammable
▶ Extremely flammable

Let's look at the definition of these categories more closely.

Flammable

This describes a product which is easily ignitable and capable of burning rapidly. Note that the word inflammable has the same meaning as flammable.

In the UK a flammable liquid is defined as a liquid that has a flash point of between 21°C and 55°C. However, in the USA there is a precise definition of flammable liquid as one with a flash point below 100°F (37.8°C).

Highly flammable

This is describing a product which has a flash point below 21°C but which is not defined as extremely flammable.

Extremely flammable

This describes a product which has a flash point lower than 0°C and a boiling point of 35°C or lower.

2.1.5 Fire triangle

For a fire to start, there are three elements which have to be present. These are:

▶ a source of fuel
▶ a source of heat or ignition
▶ oxygen

This is known as the 'fire triangle'.

If any one of these elements is eliminated, then a fire cannot continue or start. We shall be looking at the subject of fire triangle in more detail in Chapter 15.

Figure 2.2 Fire triangle
Source: Wise Global Training.

2.1.6 Flammable range

Within the fire triangle where the source of fuel is a vapour, for it to burn it has to fall within a specific percentage range in comparison with air. If the mixture of flammable vapour and air has too much flammable vapour in it, the atmosphere will not burn. This is described as an atmosphere that is 'too rich' to burn.

Conversely, if the mixture of flammable vapour and air has too little flammable vapour in it, the atmosphere will again not burn. This is described as an atmosphere that is 'too lean' to burn.

The **lower flammable limit** is the lowest concentration of a gas or vapour in air which is capable of being ignited.

The **upper flammable limit** is the highest concentration of a gas or vapour in air which is capable of being ignited.

The percentage of flammable vapour which falls between these two parameters (not too rich to burn but also not too lean to burn) is known as the **flammable range**.

For example, methane gas has a lower flammable limit of 4.4 per cent and an upper flammable limit of 15–17 per cent. This is the particular **flammable range** of methane gas, and at any point between these two limits the atmosphere is flammable.

There is always a risk of fire and explosion in an area which contains vapour within the flammable range. It is vital, therefore, to control the atmosphere to make sure that the flammable range is not reached. Purging (replacing) the air in storage tanks with nitrogen is such a control. This is because nitrogen is an inert gas (i.e. it will not burn).

2.1.7 Toxicity

Toxicity is used in two senses:

(a) To denote the capacity to cause harm to a living organism

(b) To indicate the adverse effects caused by a chemical

This information is available on the Material Safety Data Sheet which comes with any chemical.

With regards to toxicity, the following further descriptions may be used:

▶ **Acute toxicity**
▶ **Chronic toxicity**

Let's look at the definitions of these terms more closely.

Acute toxicity is a term which describes the effect a substance has had on a person after either a single exposure or from several exposures within a short space of time (e.g. 24 hours or less). An example of this would be from radiation.

Chronic toxicity is a term which describes the effects a substance has had on a person after many exposures over a longer period of time (e.g. months or years). An example of this would be exposure to asbestos.

2.1.8 Skin irritant

The Occupational Health and Safety Administration (OSHA) describes skin irritant as 'a chemical which is not corrosive, but which causes a reversible inflammatory effect on living tissue by chemical action at the site of contact'.

This means that the substance may cause a local inflammatory reaction of the skin exposed to it either immediately after one exposure or from repeated exposure. However, the inflammation is reversible.

Reactions can be strong or weak. Substances which cause a strong reaction may possibly be diluted in order to weaken the reaction.

The immediate reaction from strong irritants is called acute irritant contact dermatitis. On the other hand, it may take a number of exposures over a longer period of time for weak irritants to cause a reaction. This is known as chronic irritant contact dermatitis.

2.1.9 Carcinogenic properties

A carcinogen is defined as any substance that can cause, or aggravate, cancer. They fall into two groups:

▶ **Genotoxic carcinogens**
▶ **Non-genotoxic carcinogens**

Genotoxic carcinogens are those which react with DNA directly or with macromolecules which then react with DNA. There are no safe thresholds of exposure to genotoxic carcinogens.

Non-genotoxic carcinogens do not react directly with DNA although they do cause cancer in other ways. There may be some threshold exposure limits for substances which fall within this group.

2.2 Properties and hazards of various gases associated with the oil and gas industry

2.2.1 Hydrogen

Hydrogen is a gas which is difficult to detect as it is odourless and colourless. It is lighter than air (with a density of 0.07 when compared with air) and so will rise when released. Hydrogen is a highly flammable gas when it is mixed with air (flammable range 4–75 per cent) and it burns with an invisible flame. The only way to detect burning hydrogen is when it ignites something else which has a visible flame. Consequently, should hydrogen be suspected to be present within the atmosphere, it is essential to ensure there are no sources of ignition in the vicinity.

2.2.2 Hydrogen sulphide (H_2S)

Hydrogen sulphide (H_2S) is produced from decaying vegetation and marine micro-organisms. It is a toxic, corrosive and flammable gas. It is a hazard for the oil and gas industry as it can be released as it comes to the surface with drilling shale.

Hydrogen sulphide (H_2S) is an extremely dangerous gas. It has a density of 1.39 when compared with air and tends to drift in low lying areas such as pits, cellars, drains, etc. As such, it is difficult to disperse.

It is toxic when inhaled because, when it enters the bloodstream, it combines with the haemoglobin in red blood cells, preventing the absorption of oxygen into the blood, thus rapidly causing asphyxiation. It has an effect which is similar to that of carbon monoxide.

Although hydrogen sulphide has a foul odour, it very rapidly paralyses the sense of smell and can quickly overcome anyone exposed to it and asphyxiate them. Even very low concentrations of the gas can prove fatal. For this reason it has a Time Weighted Average (TWA) of 8 hours at 5 ppm, or 15 minutes at 10 ppm. These limits are those which are applicable in the UK as set by the UK Health and Safety Executive. However, international limits may vary.

2.2.3 Methane

Methane is an odourless, colourless gas which exists naturally in the substrate. It is lighter than air with a density of 0.717 compared with air. It is a flammable gas and, when mixed with air in concentrations between 5 and 15 per cent, is explosive.

Although methane is not toxic at low concentrations, it can cause asphyxiation if the level is high enough to reduce the amount of inhaled oxygen.

2.2.4 Liquefied Petroleum Gas (LPG)

Liquefied Petroleum Gas (LPG) is a mixture of hydrocarbon gases which are highly flammable and used as fuel in heating and cooking appliances and motor vehicles. It is also used as an aerosol propellant and refrigerant. The two main types of LPG are butane and propane.

LPG is an odourless, colourless gas which has a density of 2.0 when compared with air and tends to drift in low lying areas such as pits, cellars, drains, etc. As such, it is difficult to disperse.

Liquefied Petroleum Gas (LPG) expands at a rate of 250:1 at atmospheric pressure when it changes from a liquid to a gas. Consequently, it can cause a massive vapour cloud from a relatively small amount of liquid when that liquid is released into the air.

Another issue with liquefied petroleum gas (LPG) is that, to store it effectively, it has to be converted from a gas to a liquid, which means it is stored at a temperature of between 0°C and −44°C. Consequently, any moisture which settles to the bottom of the tank storing the LPG will need to be drained off. This operation is extremely hazardous as it carries the risk of this water freezing the drain valve in an open position and allowing LPG to escape.

This very scenario occurred in 1966 at Feyzin in France. The resulting Boiling Liquid Expanding Vapour Explosion (BLEVE) killed fifteen people and injured a further eighty-one.

Liquefied petroleum gas (LPG) is toxic and can cause:

▶ Asphyxiation
▶ Cold burns to the skin on contact
▶ Brittle fracture to carbon steel on contact
▶ Environmental damage
▶ Fire and explosion

2.2.5 Liquefied Natural Gas (LNG)

Liquefied Natural Gas (LNG) is a colourless, odourless, highly flammable natural gas which is made up of methane (85–95 per cent), ethane, propane and butane. It is non-corrosive and non-toxic although, like methane, it can cause asphyxiation if the concentration is high enough when inhaled.

To convert natural gas from a gas to a liquid it needs to go through a process of condensation using liquid nitrogen. This reduces the temperature to −162°C where it takes up 600 times less space in its liquid form than it does as a gas. The hazards associated with Liquefied Natural Gas (LNG) are:

▶ Asphyxiation
▶ Cold burns to the skin on contact
▶ Brittle fracture to carbon steel on contact
▶ Fire and explosion

2.2.6 Nitrogen

Nitrogen is the most abundant gas in the earth's atmosphere, 78 per cent by volume. It is a colourless, odourless, non-flammable gas which is often used as a blanket gas in storage tanks and for purging equipment and processes of oxygen and hydrocarbons, thus eliminating the hazards of fire and explosion.

The primary hazard associated with nitrogen is asphyxiation when it is used in confined spaces to displace oxygen.

2.2.7 Oxygen

Oxygen is an odourless, colourless gas which is present in the atmosphere. It is vital for sustaining life as it is breathed in and absorbed into the bloodstream. It is considered to be a safe gas, but it can present the following hazards:

▶ Asphyxiation – this is because the body is stimulated to breathe by the level of carbon dioxide (CO_2) in the air and should a situation arise where oxygen is

released into an area displacing the carbon dioxide, then the stimulus to breathe could cease, causing death by asphyxiation.

▶ It is one element of the 'fire triangle', i.e. it allows a fire to burn.
▶ It can oxidize metal, i.e. cause rusting. This is a very serious hazard, especially in the oil and gas industry where large amounts of the infrastructure are manufactured from carbon steel and are at risk of failure due to rusting. It is not always obvious where rusting occurs, especially if it is within areas that are impossible to see, e.g. inside hollow structures.

2.3 Properties and hazards of associated products and their control measures

2.3.1 Anti-foaming agents and anti-wetting agents

Anti-foaming and anti-wetting agents are used to prevent foam forming or to break down foam that has already been created in a process liquid during any production processes. Foam can have a detrimental effect on product quality and production efficiency by slowing down the process. Some anti-foaming agents are oil based and others are silicone based.

Anti-wetting agents are coatings which are applied to surfaces of vulnerable components which are subject to moisture and subsequent corrosive activities. They are known as 'hydrophobic' coatings, meaning they repel moisture.

Although both these types of agents are generally non-hazardous, it is advisable to wash any areas of skin contact with soap and water. Information provided by the product's Material Safety Data Sheet (MSDS) should always be consulted when they are used.

2.3.2 Micro-biocides

Micro-biocides are used to protect against the harmful effects of bacteria, e.g. legionella, which can proliferate in air conditioning systems and humidifiers. They do this by destroying the bacteria if it is present or by preventing its formation.

They are also used as corrosion inhibitors on some metals, e.g. steel pipelines.

Micro-biocides are classed as irritants to skin and eyes on contact, as well as being toxic if ingested.

2.3.3 Corrosion preventatives

Corrosion is a major hazard in the oil and gas industry. Preventing corrosive activity from causing damage to infrastructure is essential. Some corrosion preventatives come in the form of a water-displacing film which acts by spreading across the surface of metals, displacing water from cracks and crevices and forming a barrier to corrosive activity.

Other types of preventatives are applied to metals; they then dry to a hard resin or waxy film, thus forming a barrier to corrosive activity.

Information provided by the product's Material Safety Data Sheet (MSDS) should always be consulted when they are used.

2.3.4 Refrigerants

Refrigerants are liquefied gas under pressure and pose a minimal risk provided they remain contained within their allotted systems. Problems arise when there is an escape or release of the refrigerant because it can displace oxygen in the atmosphere with the potential to provoke asphyxiation. To minimize the risk it is advisable to conduct regular safety checks on all equipment and systems, and have in place emergency procedures to deal with any unexpected release of the refrigerant.

Data relating to the refrigerant will be contained in the Material Safety Data Sheet (MSDS) which is provided with the product. This should state the toxicity status of the product as well as the safe exposure limits which apply.

The hazards associated with refrigerants are:

▶ Injury from components or material ejected by the high pressure escape
▶ Frostbite injury to skin or eyes where contact with refrigerant is made
▶ Asphyxiation
▶ Possible explosion or fire if the refrigerant is flammable
▶ When certain refrigerant gases burn they can produce other gases which can be very toxic
▶ Liquid refrigerants have a very high expansion rate when changing from a liquid to a gas, causing overpressure
▶ Refrigerant gases are heavier than air and will slump if the gases are accidently released

Good practice and safe working procedures when dealing with refrigerants are as follows:

▶ Having procedures in place to deal with any unexpected release of refrigerant, e.g. recovery procedures and equipment to contain the refrigerant.
▶ Never working in confined spaces where there is a risk that refrigerants may be released. This is because of the very real risk of asphyxiation.
▶ Providing ventilation equipment to deal with any potentially high concentrations of refrigerant.
▶ Ensuring that anyone who is exposed to refrigerant gas is immediately moved out of the affected area to a place where they can breathe fresh air and be given oxygen as necessary. They will also need to be medically examined.

2.3.5 Water/steam

Water is used for cooling and dilution in process operations as well as for fighting fires, cleaning and within air conditioning systems. It does, however, present hazards, including:

▶ Legionella, which proliferates in air conditioning systems. Regular testing and maintenance systems can control this.
▶ Leptospira, which can be found in water which has its source in freshwater rivers or lakes. This can be transmitted to humans via broken skin or through the mucous membranes of the eyes, nose or mouth and, in extreme cases, can cause Weil's disease, which can be fatal. Effective personal hygiene can control this.
▶ Corrosion, which attacks steel. Controls for this include applying a protective coating to steel components or fitting sacrificial anodes within the system to provide cathodic protection.
▶ The build-up of an electrostatic charge within pipework resulting in a potential explosion if this is discharged. The electrostatic charge can be caused by the friction of water flowing through the pipework. To avoid electrostatic charge build-up, the flow rate of water should be controlled at a suitable rate and the pipework should be bonded to earth.
▶ An increase in pressure within pipework and other components within a system. This may be caused by an increase in temperature during hydrostatic pressure testing, e.g. where there is exposure to direct sunlight. This may cause failure of the system with catastrophic effects.

Freezing water

Water expands when it freezes, and this can result in the failure (fracture) of pipework and/or other components within a system. Under certain conditions, plugs of ice (hydrates) can form and these can block pipes and pumps as well as preventing the closure of valves which, in critical situations, can have catastrophic effects.

A prime example of this was the Feyzin disaster in France in 1966, which we mentioned earlier. There, an operator was draining water from a pressurized propane tank when a hydrate plug formed in the drain valve. Consequently, he was unable to close the valve and a cloud of propane vapour escaped and exploded when it came into contact with a source of ignition.

Controls preventing this can be:

▶ Lagging pipes considered to be at risk of being damaged by freezing water
▶ Fitting steam trace lines
▶ Draining unused components

Sea water

Sea water contains living organisms which can proliferate and cause blockages, e.g. in the heads of sprinkler systems. This can be avoided by the implementation of a regular maintenance programme to ensure all parts of a system are kept free from blockages, as well as using additives to kill any living organisms which may be present. A dry riser would also be a useful control.

Steam

Steam is used extensively in the oil and gas industry. It is used to power turbines and generate electricity, as well as serving as a source of heat and/or energy to assist with many other operations and processes. It can also be used to protect systems from the risk of freezing (tracer lines) and to serve as a heating system for areas where personnel are housed.

Steam has inherent risks as follows:

▶ It can cause thermal shock to a system if it is introduced into cold pipes or steam lines.
▶ It can cause failure of parts of a system if there is an uncontrolled expansion within it, e.g. jointed flanges could fail as they are some of the weakest points of the system.
▶ It can cause burns if anyone comes into contact with it.

2.3.6 Mercaptans

These are substances containing sulphur which are used to help detect natural gas by giving it an odour (natural gas in pure form is odourless). T-butyl mercaptan blends are used for this purpose as they smell of rotting cabbage, even in low concentrations in air.

Some mercaptans are harmful. For example, methyl mercaptan has the following hazards associated with it:

▶ It is harmful if inhaled.
▶ It is a respiratory irritant – chronic exposure may cause lung damage.
▶ It is a skin irritant.
▶ It is an eye irritant.
▶ It can depress the central nervous system.
▶ It has a flashpoint of −18°C.

2.3.7 Drilling muds (drilling fluid)

When drilling operations are in progress, mud is pumped from the mud pits through the 'drill string' where it is sprayed onto the drill bit. This allows for the cooling and cleaning of the drill bit throughout its operation.

It also allows the crushed rock cuttings, which have been drilled from the bore hole, to be carried to the surface where they are separated from the mud by the

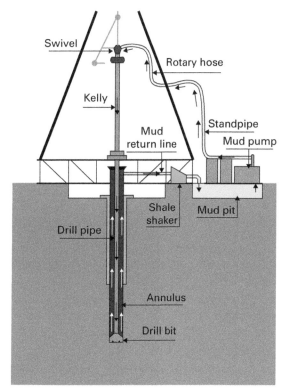

Figure 2.3 Drilling operation process
Source: Wise Global Training.

use of a 'shale shaker' or other equipment before the mud is returned to the mud pit to be reused.

Crushed rock cuttings which are too small to be separated by the shale shaker will remain in the mud and be returned to the mud pit where they can settle to the bottom. These small cuttings are called 'fines'.

Also, there may be natural gases or other flammable materials which have combined with the mud during the drilling operation. These have the ability to be released from the mud anywhere within the system where the mud is flowing back to the pit.

As a result of this there is risk of fire or explosion should these gases be exposed to a source of ignition. Control measures to prevent this include:

▶ Safe working procedures
▶ Monitoring sensors
▶ Equipment and wiring which has been certified as explosion proof

Once the mud has been returned to the mud pit it is tested and, if necessary, further additives are combined with it to give it the required properties before it is pumped back into the system in a continuous cycle.

Functions of drilling mud

The main functions of drilling mud are as follows:

▶ Remove cuttings from the well
▶ Suspend and release cuttings
▶ Control formation pressures
▶ Seal permeable formations
▶ Maintain wellbore stability (line the walls of the hole)

▶ Minimize formation damage
▶ Cool, lubricate and support the bit and drilling assembly
▶ Transmit hydraulic energy to tools and bit
▶ Ensure adequate formation evaluation
▶ Control corrosion (to an acceptable level)
▶ Facilitate cementing and completion

There are different types of drilling fluids: water based, oil based and synthetic based muds.

We will now look in more detail at these three types of drilling fluid.

Drilling muds – Water Based Mud (WBM)

Water Based Mud (WBM) is a combination of clay and other additives blended with water to make a thick fluid. The more additives in the water, the thicker the mud will be. The fluid is normally made from indigenous clays although should particular muds be required the additives may be brought in from specialist suppliers.

Bentonite is a common additive which allows the mud to be fluid and free flowing when it is being pumped into the system and becomes semi-solid or gel-like when the pumping stops. Once pumping resumes it reverts to being free flowing.

Another additive which can be used is potassium formate, which allows the mud to have various other characteristics. These include:

▶ Viscosity control
▶ Shale stability
▶ Enhancing the drilling rate of penetration
▶ The cooling and lubricating of equipment

Drilling muds – Oil Based Mud (OBM)

Oil Based Muds (OBMs) have, as their name suggests, oil – usually diesel oil – as their base fluid. The advantages oil based muds bring to the drilling process include:

▶ Increasing lubrication of the drill shaft
▶ Enhancing shale inhibition
▶ Adding greater cleaning ability

OBMs allow for higher working temperatures to be used without adverse effects. However, there are environmental considerations to be taken into account when considering the use of OBMs.

Drilling muds – Synthetic Based Muds (SBMs)

These muds have the properties of oil based muds but have the advantage of being less toxic as their base is made from synthetic oil.

2.3.8 Low Specific Activity (LSA) sludges

The formations of rock and shale which contain oil and gas deposits also contain Naturally Occurring Radioactive Materials (NORM). These include:

▶ Uranium
▶ Thorium
▶ Radium
▶ Lead-210

Oil and gas were created in the earth's crust by the decay of sea life in ancient seas and are, therefore, often found in aquifers which contain salt water (brine). Various minerals, as well as radioactive elements, are also dissolved in the brine and these separate out and form wastes at the surface. These include:

▶ Mineral scales inside pipes
▶ Sludges
▶ Contaminated equipment or components
▶ Produced waters

Workplace equipment where sludges might be found

The process of extraction exposes the environment and humans to the radioactive elements in the sludges. As such, they are classified as hazardous. They can be found in the following locations:

▶ On the drill string
▶ Inside vessels (demister pads)
▶ Inside filters
▶ In coalescars (coarse filter/emulsifier)
▶ In coolers where tubes might be coated with sludge

As we've mentioned, sludges are a mixture of liquid and suspended material and therefore present a range of hazards, for example:

▶ Skin irritant (possibly causing dermatitis)
▶ Inhalation (fumes or dust from dried sludges)
▶ Ingestion (poor hygiene, i.e. not washing, not cleaning up, eating at work site)
▶ Radiation
▶ Carcinogenic
▶ Environmental (pollutant)
▶ Absorption through the skin (dermatitis)

The main hazard of exposure to ionizing radiation from Low Specific Activity (LSA) materials is that of the inhalation and ingestion of radionuclides, especially of dust and fumes. Employees are at a higher risk of significant exposure to ionizing radiation if they work with dusty processes unless adequate control measures have been put in place to prevent the inhalation of dust.

Employees may be exposed, although to a lesser extent, to direct radiation where there is bulk storage of the material. They may also be exposed to external radiation if they are involved in cleaning operations or the dismantling of equipment which contains scale from oil and gas extractions.

Control measures that reduce risk to workers exposed to sludge

Employers should put in place controls to ensure that the risk of exposure to ionizing radiation is reduced 'so far as is reasonably practicable'. The use of personal protective equipment may be used but only as a last resort after all other control measures have been considered. Engineering controls and the implementation of safe systems of work should take priority, and they include:

▶ The provision of ventilation equipment to contain dusts and fumes

▶ The use of wet methods of working and good housekeeping to reduce the amount of dust in the atmosphere

▶ Having equipment in place to collect sludge instead of using manual means

▶ Diluting sludge with water

▶ The use of permit-to-work systems – especially if the concentrations of dust or fumes reach a level where only designated or classified persons are allowed to work under the restrictions of a written safe system of work

▶ The provision of training and awareness programmes

▶ The provision of a health surveillance programme to monitor the health of employees

▶ The use of Respiratory Protective Equipment (RPE) specifically chosen to protect against exposure to airborne radioactivity

REVISION QUESTIONS FOR ELEMENT 1 CONTINUED

4 **Explain** the difference between 'vapour pressure' and 'vapour density'.

5 **Explain** why, when considering the hazards associated with a vapour or gas, the density of the vapour or gas is important.

6 (a) **Identify FOUR** hazards associated with refrigerants.

   (b) **Give TWO** control measures relating to refrigerants.

You can compare your answers with our guidance at the end of the book under the section entitled 'Revision and assessment guide'.

CHAPTER 3

Element 1

Sub-element 1.3:
Risk management techniques used in the oil and gas industries

Chapter contents

3.1 The purpose and uses of risk assessment techniques, qualitative and quantitative techniques

3.1.1 Introduction

Within the oil and gas industry there are inherent risks of accidents occurring at any stage of the process – from exploration through to the extraction, refining and final delivery of the product. These risks include fire, explosion, environmental contamination and injury to personnel.

There is a responsibility within the oil and gas industry to identify those risks and put in place control measures to reduce them to a level that is as low as is reasonably practicable. In order to do this, resources need to be made available to ensure that those control measures are robust and appropriate, and that the industry is staffed by well-trained people who have the experience and knowledge necessary in order to perform their work safely and effectively.

3.1.2 What is a risk assessment and what is its purpose?

To put in place risk control measures it is important to identify those risks, the first step of which is to perform a risk assessment. This allows those risks which are relevant to be identified and be given appropriate consideration. In the oil and gas industry, those risks are generally associated with all plant, equipment, products, processes and systems of work, all of which have the potential to cause harm.

There are a number of techniques available when assessing risks, including:

▶ The 5-step approach
▶ Qualitative assessment techniques
▶ Semi-quantitative assessment techniques
▶ Quantitative assessment techniques

We shall be looking at each of these in more detail as we progress. However, let's just remind ourselves of what a hazard and risk are.

A **hazard** is defined as something with the potential to cause:

▶ Harm including ill health and injury
▶ Damage to property, plant, products or the environment
▶ Production losses or increased liabilities

A **risk** is defined as the likelihood that the harm will occur: the chance – high or low – that somebody could be harmed or some infrastructure could be damaged. This is usually accompanied by an indication of how serious the harm could be.

Now we need to look at the four general risk assessment techniques we just mentioned.

3.1.3 The 5-step approach to risk assessment

Step 1 Identify the hazards
Step 2 Decide who might be harmed and how
Step 3 Evaluate the risks and decide on precautions
Step 4 Record the findings and implement them
Step 5 Review the assessment on a regular basis and update if necessary

Let's look at each of these steps in more detail to understand what they mean in practice.

The 5-step approach – step 1: identify the hazards

The first step is to work out how people could be harmed. In order to help identify the hazards the following procedure should be followed:

▶ Conduct a tour of the workplace and observe what could reasonably be expected to cause harm.
▶ Consult the workers or their representatives for their views and opinions.
▶ Consult the manufacturers' instructions or data sheets. These will highlight hazards associated with machinery or substances.

▶ Consult the accident log and ill-health records. These can often indicate less obvious hazards as well as highlighting trends.

The 5-step approach – step 2: decide who might be harmed and how

For each hazard, there has to be clear identification of the groups of people who might be harmed – this will help identify the best way of managing the risk (e.g. 'people working in the boiler room' or 'passers-by'). In each case, identify how they might be harmed, i.e. what type of injury or ill health might occur. For example, workers lifting heavy equipment may be susceptible to back injuries.

The 5-step approach – step 3: evaluate the risks and decide on precautions

Having identified the hazards, the next step is to decide what action to take to reduce the risks associated with the hazards. In most countries the law requires employers to do everything 'reasonably practicable' to protect people from harm.

Setting controls

When setting controls to minimize the risks to As Low As Reasonably Practicable (ALARP), the **Hierarchy of Control** should be used. We shall be looking in more depth at ALARP later in this chapter.

When using the Hierarchy of Control, priority should be given to those control measures at the top of the list.

▶ **Elimination**
▶ **Substitution**
▶ **Engineering controls**
▶ **Administrative controls**
▶ **Personal Protective Equipment (PPE)**

The 5-step approach – step 4: record the findings and implement them

Implementing the results of the risk assessment is the next step. The first step of implementation is to write down the results of the risk assessment and share the document with those staff members involved.

A risk assessment is not expected to eliminate all risks, but it is expected to be suitable and sufficient. In order for it to meet these criteria, it will need to be able to show that:

▶ A proper check was made.
▶ All of those who might be affected were consulted.
▶ All the significant hazards were addressed.
▶ The recommended risk control measures are suitable and sufficient, and the remaining risk is low.
▶ All the staff or their representatives were involved in the process.

If the findings of the risk assessment conclude that there are a number of improvements to be made, it is appropriate to draw up a prioritized plan of action.

The 5-step approach – step 5: review the risk assessment and update if necessary

Few workplaces remain static. Inevitably, new equipment or variations in substances used and procedures undertaken will introduce new hazards to the workplace. Consequently, it's sensible to review all control measures on an ongoing basis.

Over and above the 5-step approach to risk management, there are other techniques available which take a different approach to risk assessment and control. These include **qualitative risk assessment techniques, semi-quantitative risk assessment techniques** and **quantitative risk assessment techniques**. These techniques are regarded as more comprehensive and can be used to take in a wider range of factors including financial costs, loss of time, loss of business, loss of reputation, etc.

Let's take a look at these techniques in terms of the opportunities they provide as well as their limitations.

3.1.4 Qualitative risk assessment

A qualitative risk assessment is based on the conclusions reached by the assessor using his/her expert knowledge and experience to judge whether current risk control measures are effective and adequate, in order to ensure they reduce the risk to a level which is as low as is reasonably practicable, or if more measures need to be applied.

It's a way of identifying hazards emanating from specific activities which might affect people or the environment. The assessor can develop an understanding of the risks involved and how serious they may be if realized, thus allowing him/her to prioritize the control measures in the order that they should be implemented. The use of a scale matrix may be helpful in this process.

There are advantages in using the combined skills of a team of assessors. Getting a fuller, more rounded picture of the risks involved would result from having a pool of ideas and judgements rather than those of a single assessor. Where a team of assessors is involved, having them work independently on risk assessment at the outset before bringing them together will overcome undue influences from stronger members of the team. This would conclude with a debate and comparison of ideas in order to reach a consensus of opinion and a final decision on which risk control measures should be applied.

When making a qualitative judgement on the severity of a risk, two parameters are taken into consideration. These are the likelihood of an event occurring and the consequences or severity if the event does occur. Severity can be assessed in terms of its effect on:

▶ Harm caused
▶ Time
▶ Cost

23

▶ Quality
▶ Inconvenience

3.1.5 Semi-quantitative risk assessment

Effective risk management uses well-founded decisions based on as broad a knowledge base as possible, i.e. the knowledge and experience of the assessor(s). It also requires a degree of consistency in making judgements. Under qualitative assessment techniques, both the likelihood and the severity of any event are subjective (i.e. a personal opinion). However, using a semi-quantitative approach involves putting a value on the likelihood and severity of an event. To do this effectively, a numerical value is applied to the degree of severity as well as the likelihood of a particular event occurring. An example of the kind of rating used, where measures ranging from 1 to 5 are applied, is given in Table 3.1.

Table 3.1 Numerical values applied to levels of likelihood and consequence

Likelihood can be defined as	Severity can be defined as
5 Very likely	5 Catastrophic
4 Likely	4 Major
3 Fairly likely	3 Moderate
2 Unlikely	2 Minor
1 Very unlikely	1 Insignificant

Source: Adapted from www.hse.gov.uk/quarries/education/overheads/topic5.doc by David Mercer.

Semi-quantitative risk assessment – risk rating/prioritization

When judging the risk of a particular activity, the risk assessor or risk assessment team agree the likelihood rating, e.g. 3, agree the consequence (severity) rating, e.g. 4, then multiply the likelihood (3) by the consequence (4) to get a rating of 3 × 4 = 12 (tolerable). This can be seen in the matrix shown in Figure 3.1 below.

This form of semi-quantitative risk rating system gives an overall numerical value to the risk being evaluated. That numerical value can then be used to prioritize the actions required, as shown in the grading on the right of the matrix.

Semi-quantitative risk assessments also offer a degree of consistency. By using a matrix like this, trained risk assessment teams across the organization are more likely to develop a more consistent approach to risk rating.

3.1.6 Quantitative risk assessment

In the oil and gas industry, the hazards associated with complex processes and operations require a sophisticated approach in order to evaluate the risks involved. Quantitative risk assessment techniques provide the means to make a detailed assessment that will be based on quantitative considerations of event probabilities and consequences.

The quantitative risk assessments will involve using special quantitative tools and techniques in order to identify hazards, to give an estimate of the severity of the consequences and the likelihood of the hazards being realized. The quantitative risk assessments will result in the provision of numerical estimates of the risks, and these can then be evaluated when compared with known numerical risk criteria.

Historical data analysis

Historical data analysis is the basis for many quantitative risk assessments. Frequencies are simply calculated by combining accident experience and population exposure, typically measured in terms of installation-years:

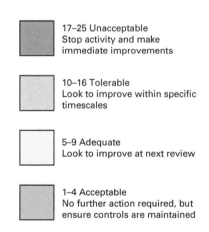

Figure 3.1 Risk rating matrix
Source: Wise Global Training.

$$\frac{\text{number of events}}{\text{number of installations} \times \text{years of exposure}}$$

One example of a source of historical data which can be used as the basis for quantitative risk assessments is the Worldwide Offshore Accident Databank (WOAD).

3.2 How risk management tools are applied in process safety risk identification and assessment, application in project phases from concept, design, start up, the concept of ALARP and the management of major incident risks

---

3.2.1 Introduction

When a project is in the design stage, some risks can be 'designed out', as can some of the hazards, using modelling as a tool. Unfortunately, the hydrocarbon inventory will always remain as a major hazard, as it's the very reason the industry is there in the first place.

Examples of modelling techniques are:

▶ HAZID (Hazard Identification Study)
▶ HAZOP (Hazard and Operability Study)
▶ FMECA (Failure Modes and Effective Critical Analysis)/FMEA (Failure Modes and Effects Analysis)

We are now going to look at each of these modelling techniques in more detail.

3.2.2 HAZID (Hazard Identification Study)

A Hazard Identification Study (HAZID) is, as its name suggests, a tool for identifying hazards. It is normally a qualitative risk assessment and is judgement based. It is usually undertaken by a team of people who will be selected because of their particular knowledge, experience or expertise.

The reasons for identifying hazards are twofold, as follows:

1 To compile a list of hazards which can then be evaluated using further risk assessment techniques. This may be described as 'failure case selection'.
2 To conduct a qualitative evaluation of how significant the hazards are and how to reduce the risks associated with them. This may be described as 'hazard assessment'.

The following features are essential elements of a hazard identification study:

▶ The study should be creative and dynamic. This will allow a wide scope of hazards to be considered.

▶ The study should take a structured approach so as to be comprehensive in its coverage of relevant hazards.
▶ The study should embrace historical data and previous experiences so that lessons learned can be acted upon.
▶ The scope of the study should be clearly defined. This is to ensure that those who read the study fully understand which hazards have been included and which have been excluded.

Hazard checklists

These are an effective means of producing a comprehensive list of standard hazards which can be used for hazard identification studies at the concept and design stages of a project to consider a wide range of issues related to safety. It is also used to confirm that good practice has been built into a project at the design stage.

The use of keywords as a prompt can be useful when considering hazards in a HAZID. Table 3.2 shows examples of keywords and some of their associated hazards. The list is not intended to be comprehensive.

Table 3.2 Hazard checklist

Key words	Hazards
Fire	Blowout which has ignited
	Process leak which has ignited
	Product storage leak which has ignited
Loss of breathable atmosphere	Ingress of smoke
	Asphyxiation
Toxic gas release	Ingress of toxic gas
	Asphyxiation
LPG/LNG release/leak	Explosion from contact with a source of ignition
	Hydrate formation on valves
	Cold burns/frostbite
	Brittle fracture of steel component(s)
Hydrocarbon release/ leak	Explosion from contact with a source of ignition
Collision/crash	Helicopter crash
	Vessel colliding with rig
Structure failure	Crane collapse Rig leg collapse

Source: Adapted from New South Wales Department of Planning (2011) *Hazardous Industry Planning Advisory Paper No. 8.* HAZOP Guidelines, Sydney, State of New South Wales ISBN 978 0 73475 872 9 available at: http://www.planning.nsw.gov.au/Portals/0/HIPAP%208%20Final%202011.pdf

The strengths of a hazard checklist are that:

▶ It is relatively cheap to produce and can be created by a single analyst.
▶ It can be used to help prevent the recurrence of previous incidents.
▶ It can be used for concept designs with a minimum of installation information.

▶ It can use the experience gained from previous risk assessments.

Its weaknesses are that:

▶ It may not be able to anticipate accidents which may occur in new designs.

▶ Using a generic checklist does not encourage new thinking about possible hazards, which can limit the understanding of the types of hazard specific to the installation.

In conclusion, a generic checklist is a useful tool for most risk assessments, although it is advisable to use it alongside other hazard identification study methods.

3.2.3 HAZOP (Hazard and Operability Study)

Hazard and operability studies are usually conducted at the design stage of a plant in general and the operating and safety control systems in particular. Consequently, they offer an opportunity to pre-empt hazards at a stage where they can be evaluated and dealt with before they become a reality.

A Hazard and Operability Study (HAZOP) is a tool which is used to systematically examine every part of a process or operation in order to find out how deviations from the normally intended operation of a process can happen, and if further control measures are required to prevent the hazards, which have evolved from the study, from happening. In order to do this, the HAZOP uses a complete description of the process, including Process and Instrumentation Diagrams (P&IDs) or their equivalent.

The hazard and operability study uses a qualitative assessment technique to ask 'what if' questions to identify problems before the start of operations. Every part of the installation is systematically examined by a team which comprises experts with a wide range of skills and experience relevant to the installation. The costs of conducting the HAZOP and any recommended implementations will be more than offset by the savings in resources – from commissioning times to lives saved.

The questions are set around guide words which are developed from method study techniques. This allows the questions asked to explore every possible way the operation could deviate from the normal intended operation of the process, and thus test its integrity.

The systematic approach of this technique is advantageous in failure case identification.

The hazard and operability study is useful in the communication between the design team and the operator(s) of the installation. It also provides opportunities for the provision of training for key production staff of new installations.

As we've mentioned, HAZOPs may be used at the design stage, but they can also be used when plant alterations or extensions are to be made or applied to an existing facility.

The HAZOP procedure involves selecting a line in the process, and one of the team, with the appropriate knowledge, describes the normal operating procedure or function of this line. Various scenarios, prompted by the guide word list, such as **HIGH PRESSURE** are then applied. Consideration is first given as to what could cause this particular deviation. Thereafter, the consequences or results of the deviation are discussed.

The next step is for the team to consider how credible this particular scenario is, whether its effects are significant, and whether additional safeguards are required.

Let's now look at an example of the logical sequence of steps in conducting a hazard and operability study.

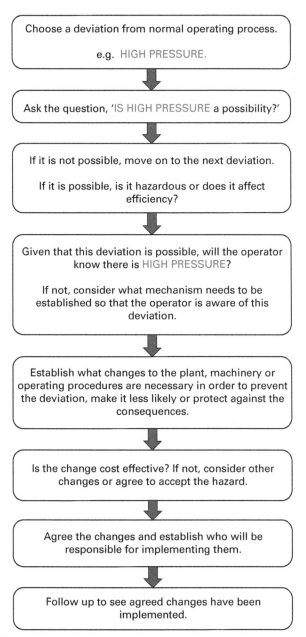

Figure 3.2 HAZOP study flow chart
Source: Wise Global Training.

Hazard and Operability Study (HAZOP) – process guide words

In our example we used **HIGH PRESSURE** as the deviation. The following are other examples of guide words used in a HAZOP analysis. This list is not meant to be exhaustive.

Table 3.3 HAZOP study guide words

FLOW – HIGH	LOW	ZERO	REVERSE
LEVEL – HIGH	LOW		
PRESSURE – HIGH			
TEMPERATURE – HIGH	LOW		
CONTAMINATION			

Source: Adapted from New South Wales Department of Planning (2011) *Hazardous Industry Planning Advisory Paper No. 8.* HAZOP Guidelines, Sydney, State of New South Wales ISBN 978 0 73475 872 9 available at: http://www.planning.nsw.gov.au/Portals/0/HIPAP%208%20Final%202011.pdf

The effectiveness of a hazard and operability study will depend on:

- How the team is made up – the range of skills and experience of the individual members.
- The team leadership – the chairperson should ensure that the team is aware of, and follows, stringent procedures which systematically test the integrity of the design.
- How current the information is – the team should have the most up-to-date and accurate data available to them, including process and instrumentation diagrams (P&IDs).
- The team systematically examining the information to establish the cause and effects of any potential hazards resulting from deviations from the design.

Hazard and Operability Study (HAZOP) – team members

The HAZOP team should consist of a group of people who, between them, have expert knowledge in every area of the process plant and its operations. Typically they are a group of between five to eight people in the fields of management and engineering. They need to be aware of all of the details in the process and instrumentation diagrams (P&IDs) if it is a new process at the design stage but, if the plant is already in existence, then the team also needs to include process and maintenance staff to ensure that all aspects of the process plant and operations are considered. The leader of the team, the chairperson, should be fully au fait in hazard and operability study techniques in order for him/her to ensure that the team follows the procedure comprehensively and systematically.

An example of a HAZOP team that has been assembled to consider the design of a new chemical plant could comprise the following people:

- Chairperson – This should be a person who has not been directly involved with the design of the plant, but who has experience in hazard and operability studies. This is so he/she can work independently to ensure that all procedures are followed correctly. Although independent of the design team, the chairperson would benefit from having an understanding of the plant design.
- Design engineer – This is the person who has been involved with this particular project and will be available to provide information about details of the design.
- Process engineer – In this case, this will be a chemical engineer who will be the person responsible for developing the process and instrumentation diagrams (P&IDs) as well as the process flow diagram.
- Electrical engineer – The person responsible for developing the design of the electrical systems within the plant.
- Instrument engineer – This is the person who was concerned with the design and selection of the control systems for the plant.
- Operations manager – The operations manager will be the person in charge during the commissioning and operation phases of the plant.

The strengths of HAZOP are:

- HAZOP studies are well known and widely used so the advantages and disadvantages are well recognized.
- HAZOP studies use the knowledge and experience of operational staff within the team.
- HAZOP studies are used to systematically examine every part of the design, in order to identify every conceivable deviation.
- HAZOP studies can be used to identify possible technical faults as well as any human errors which may occur.
- HAZOP studies, whilst identifying existing safeguards, are also able to evolve and develop further controls or safeguards.
- The use of a team for HAZOP studies in offshore operations is advantageous in that it comprises of a wide range of disciplines from a variety of differing organizations.

Its weaknesses are:

- Its success depends on the effectiveness of the chairperson and the knowledge and experience of the team.
- It is best suited for use in identifying process hazards; for it to be used for other types of hazards it will need modifying.
- Procedural descriptions are needed, and these may not be available in sufficient depth of detail to involve all conceivable scenarios.
- The documentation required to record the study comprehensively can be extensive and overwhelming.

3.2.4 FMECA (Failure Modes and Effects and Criticality Analysis)

An FMECA (Failure Modes and Effects and Criticality Analysis) is a method of systematically identifying the failure modes of an electrical or mechanical system. One or two people examine each component of the system in turn and evaluate the effects and the degree of importance if that component should fail.

The examination uses a document that contains a systematic list of all of the components, and usually includes:

▶ The name of the component
▶ The function of the component
▶ The possible types of failure
▶ The causes of each type of failure
▶ How each failure is detected
▶ The effects of each failure on the primary system
▶ The effects of failure on other components
▶ Necessary actions to prevent each failure or what actions are necessary to repair each failure
▶ Degree of criticality

The strengths of an FMECA are:

▶ It is a well-used and understood hazard analysis tool.
▶ It may only take one person to perform the analysis.
▶ It should identify all conceivable electrical and/or mechanical hazards.
▶ It identifies safety-critical equipment/components where a failure would be critical for the system.

Its weaknesses are:

▶ It's dependent upon the experience and knowledge of the analyst.
▶ The analyst needs to develop a hierarchical system drawing before he/she can perform the analysis.
▶ It is limited to mechanical and electrical equipment and is not applicable to procedures or process equipment.
▶ Human errors and multiple failures are difficult aspects to cover.
▶ It is likely to produce a complex list of failures.

FMECAs, although useful for safety-critical mechanical and electrical equipment, should not be used in isolation. This is because human error is a contributing factor in many accidents and it is difficult for failure modes and effects and criticality analysis to identify this.

3.2.5 As Low As Reasonably Practicable (ALARP)

In this chapter, when we have been discussing the application of risk controls, we have mentioned on a number of occasions the phrase 'as low as reasonably practicable'. We now need to understand just what this phrase means.

There is risk in every aspect of our lives – in everyday activities (e.g. crossing a busy road) as well as in our working lives (e.g. slips and trips at the office or more serious accidents with machinery or working with hazardous substances). What is necessary is to reduce those risks to an acceptable level; another way of putting it is to reduce the risk to 'As Low As Reasonably Practicable' (ALARP).

What this means is that employers should adopt appropriate safety measures unless the cost (in terms of money, time or trouble) is grossly disproportionate to the risk reduction. Once all such measures have been adopted, the risks are said to be 'as low as reasonably practicable'.

Examples of the extremes might be that to spend £1 million on replacing chairs in the control room with ones with better lumbar support could be regarded as disproportionate. However, to spend £1 million on installing a fully protected escape route from the temporary refuge facility to the lifeboats and helideck could be regarded as far more proportionate.

In the oil and gas industry, the risk of fire and explosion and their consequences rank highly, not only in financial terms, but also to human life and the environment. It is because of these potential consequences that what is regarded as 'reasonable' in the oil and gas industry is at a much higher level than in most other industries. Consequently, more stringent control measures will need to be put in place to reduce the risk to a level which can be regarded as low as reasonably practicable.

3.2.6 Management of major incident risks

When it comes to managing the risks of a major incident, this should take a hierarchical approach. The recommended hierarchy is:

1 Elimination and minimization of hazards (designing safety into process and systems)
2 Prevention (the reduction of the likelihood of a major incident)
3 Detection (the warning and alarm systems transmitted to the control area)
4 Control (the limitation of the scale, the intensity and/or the duration of an incident)
5 Mitigation of consequences (the protection from effects of an incident)

Inherently safer design and measures to prevent and control major accident hazards warrant the highest priority. This is because they have the greater effect and, as such, offer dependability in reducing risk. The optimum point in time to identify and eliminate or reduce the risk of major hazards on a new installation is at the design stage. This is the stage at which all elements of the process and plant are examined and tested and where the risks are prioritized in order of significance. It is always best to prevent or eliminate risks by engineering design which will make the installation inherently safe, and then any residual risks can be controlled by the implementation of management and other controls. It is more difficult to eliminate or prevent risks on existing installations, although to comply with likely legislative requirements they should be reduced to As Low As Reasonably Practicable (ALARP).

The safety case/report of an installation should consider what the effects of fire and explosion would be on the integrity of the installation – especially offshore, i.e. there is the risk of the installation becoming unstable and unable to keep on station. There is also the risk to the environment from the effects that a release of toxic gas would have.

The lessons learned from investigations into previous incidents cannot be overstated when it comes to managing the risks associated with major incidents. An example of this is the report forthcoming from the Buncefield incident which occurred in December 2005. In this incident there was an explosion and the fires that followed it had a massive impact, not only on the plant, but also on the residents and business community locally and further afield. It was fortunate, however, that no lives were lost in this incident.

Although the recommendations which resulted from the subsequent enquiry into the incident were associated with how to respond to land-based incidents, some of the general principles can be applied to offshore facilities.

The principal recommendations were as follows:

▶ To review emergency arrangements to cover all reasonably foreseeable emergency situations which may result from credible major incidents. The word 'credible' in this instance to be broadened so that some scenarios, which may previously have been regarded as unrealistic, should be considered.

▶ To ensure that guidance, which is related to existing emergency plans, is reviewed by an external, independent authority.

▶ To ensure training is given to all relevant staff in order that they become competent in the implementation of the emergency plan, should such a situation arise. They must also ensure that there is an adequate level of trained staff available at all times.

▶ To ensure that the control centre used for an emergency situation is appropriately sited and adequately protected should an emergency situation arise, as should any facilities required for the emergency response. If any changes are needed to achieve this recommendation then the safety case/ report should be updated to show the changes.

▶ To ensure all critical emergency response resources are identified and contingency plans exist in case any of them fail.

▶ To ensure that adequate arrangements with all external emergency services have been made so that, should an emergency situation arise, they are fully prepared to deal with it.

▶ To ensure that there is regular communication between the operator of the installation and any external agencies that may be affected by it. An example of this would be the coastguard, who may be affected by the activities of the operators of an offshore installation.

3.3 Industry related process safety standards, inherent safe and risk based design concepts, engineering codes and good practice

3.3.1 Industry related process standards

The oil and gas industry is a multi-national operation and is governed by both national and international health and safety regulations and codes of practice which are developed and enforced by government departments and other authorities throughout the world (e.g. the Occupational Safety and Health Administration (OSHA), the Health and Safety Executive (HSE), etc.). Governments and enforcing authorities tend to work in conjunction with the oil and gas industry in order to develop these codes of practice and legislation, which develop out of the specialized knowledge and experience which these industries have in managing their risks. Models of risk management which have developed over time are shared globally throughout the industry and minimum standards of health and safety are commonly accepted and adopted by the multi-national companies.

The oil and gas industry has also set up a number of bodies which provide a forum for the exchange of ideas as well as a means of notification of hazards or the development of improved working practices.

3.3.2 Inherent safe and risk based design concepts

One of the main elements of developing inherently safe processes is to recognize that, by reducing the complexity of the plant at the design stage and simplifying the operation process, a significant reduction in the likelihood of accidents occurring can be achieved. This is because there is less equipment to malfunction and fewer opportunities for human error.

The design of a process which is as inherently safe as possible is the main goal of process designers. It is impossible to design out all risks, but process designers can use a hierarchical structure, with hazard avoidance being the priority, followed by the control of any risks remaining.

Control features, such as designing a system which can withstand the maximum likely pressure possible, are desirable elements where hazards cannot be designed out completely. However, where control is not possible, then mitigation by designing in means of reducing the magnitude of a hazard if it is realized is acceptable.

There are a number of principles for achieving an inherent safe design:

- ▶ Minimizing the amount of hazardous material present at any one time
- ▶ Replacing hazardous materials with less hazardous materials
- ▶ Moderating the effect a material or process might have (reduce temperature or pressure)
- ▶ Simplifying the design by designing out problems rather than adding features to deal with problems
- ▶ Designing in tolerance levels to cope with faults or deviations
- ▶ Limiting the effects of any adverse event, e.g. via bunds around storage tanks
- ▶ Allowing for human error by designing in failsafe features such as valves which fail to a SHUT position

3.4 The concept of hazard realization

3.4.1 Introduction

Hazard realization is when a system of hazard controls breaks down or fails, which in turn causes a hazardous event to occur. The realization of a hazard can be catastrophic in the oil and gas industry, and a prime example would be a Boiling Liquid Expanding Vapour Explosion (BLEVE).

We mentioned before the 1966 Feyzin disaster in France. It's worth looking in detail at what exactly happened in this incident.

Three operators were draining water from a sphere containing Liquefied Petroleum Gas (LPG). This was a routine procedure, but none of the operators undertaking the task on this occasion had any experience in performing the task. There was a written procedure which showed how to open the valves in a specific sequence, but this was not referred to by the operators. Consequently the valves were not opened in the right order (the closest one to the sphere was opened first) and an ice plug (hydrate) formed around the internal mechanism of the valve, making it inoperable.

LPG then began to flow from the drain valve, expanding and causing a huge vapour cloud. Being unable to close the valves, the operators fled the scene. The vapour cloud drifted over a nearby autoroute and ignited (possibly from a vehicle's exhaust). The flames then tracked back to the leaking sphere and resulted in a jet fire which then spread to another LPG sphere. This caused the Boiling Liquid Expanding Vapour Explosion (BLEVE) and the sphere exploded. Other spheres collapsed as their support legs buckled because of the heat which caused them to also explode.

Ultimately, five spheres were lost in the disaster even though emergency services were on scene. This was because they failed to cool adjacent tanks and the surrounding area and a domino effect was allowed to evolve.

3.4.2 Lessons learned

This particular incident raised a number of issues, including:

- ▶ The design of spheres (including insulated pipework and valves, insulation of tank supports)
- ▶ Procedures for draining water from spheres
- ▶ The need for fully trained and competent operators to do the work
- ▶ The need for emergency planning and procedures

This illustrates the need for well-trained staff to do their work in accordance with set procedures in order to reduce the risk of a hazard being realized. Operators should also be trained to deal with abnormal situations, should they arise, so they can comprehend alternative solutions if others fail.

3.5 The concept of risk control using barrier models

3.5.1 Barrier modelling

A barrier is described as something which is placed between a person and a hazard to prevent that person from being harmed. For example, a barrier around a hole in the pavement made by workmen to repair pipes is provided to prevent people from falling into that hole. Another example is barrier cream, which is applied to protect hands from harmful substances such as cleaning fluids which would cause skin irritation.

Barriers can also be intangible things which protect people from harm, e.g. knowledge and training. An example of this would be learning how to use an industrial machine safely so as to protect the operator from injury.

Another type of barrier could be the use of Personal Protective Equipment (PPE), but there is no guarantee that this will be sufficient on its own to protect the wearer from harm. PPE can be breached in a number of ways, as the model in Figure 3.3 illustrates. The barrier is designed to protect the wearer from the hazard but, as illustrated, it can fail to do this in a number of ways. Knowing how these failures happen offers an opportunity to anticipate them and install control measures to minimize the risk of harm being done.

Let's now look at what barriers we might put in place in order to reduce the potential exposure.

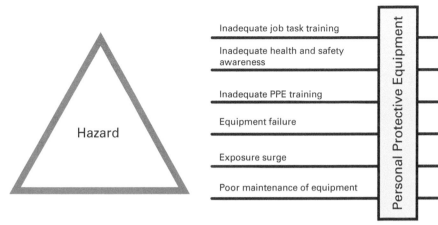

Figure 3.3 Ineffective barrier
Source: Wise Global Training.

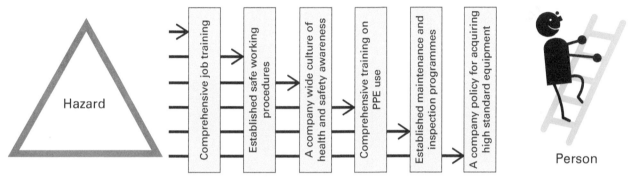

Figure 3.4 Effective barriers
Source: Wise Global Training.

3.5.2 Control measures in place to mitigate potential exposure

As shown in Figure 3.4, the risk can be minimized by putting in several barriers. Although every barrier can be breached in a number of ways, each of them reduces the exposure to some extent. In order for the hazard to be realized it would be necessary for all of the barriers to be breached simultaneously.

Examples of good barriers are:

▶ Good design and specifications
▶ Good processes and procedures
▶ Robust inspection and maintenance techniques
▶ Adequately trained and competent personnel

3.6 Use of modelling such as thermal radiation output, blast zones for risk identification

3.6.1 Introduction

There are a number of modelling tools available to help in identifying risks of fire and explosion.

The effect of heat on humans is based on a simple set of rules using kilowatts per square metre (kWm^2). At 5 kWm^2 escape from the effects of the heat can be expected. At 12.5 kWm^2 death can be expected within minutes. At 37.5 kWm^2 death is instantaneous.

The risks involved depend on the type of fire and its individual characteristics, so it is useful to look at all relevant types of fire and the modelling techniques associated with them in order to help in risk identification and subsequent risk reduction.

3.6.2 Pool fires

A pool fire is a fire burning above a horizontal and stable pool of vaporizing hydrocarbon fuel. If the fuel is not stable, it is known as a running fire.

Figure 3.5 shows a typical pool fire.

Figure 3.5 Pool fire
Source: iStock.

Pool fires present significant risks, to offshore installations in particular, which can quickly escalate into major incidents.

Modelling involves knowing the pattern of events when a pool fire takes place and being able to design in certain mitigating factors. For example, the following is a section of a Health and Safety Executive (HSE) information sheet, 'Modelling of Pool Fires in Offshore Hazard Assessments Offshore – Information Sheet no. 9/2008'.

> Liquid spills will expand on a surface until they achieve a certain critical thickness. For non-porous and relatively smooth surfaces, such as steel decks, a typical pool depth will be 1 mm for non-viscous liquids. On this basis, each square metre of pool will hold 1,000 cm^3 of liquid. If the spill of escaping liquid is ignited, the rate of pool spread is no longer simply a function of rate of fuel input, but is now governed by the balance of fuel input versus fuel burn-off rate.
>
> As the pool increases in area, the proportion of fuel burning off increases until it eventually matches the rate of input. At this time the pool should remain constant in size.

This is only one sample characteristic from the document, but it does demonstrate that knowing about this characteristic can be extremely useful at the design stage of certain components or processes.

3.6.3 Jet fires

A jet or spray fire is a flame which is being fed by hydrocarbons continuously being released with significant momentum in a particular direction.

Figure 3.6 shows a jet fire emanating from a storage tank.

Figure 3.6 Jet fire
Source: Dreamstime.

Like pool fires, jet fires present significant risks to offshore installations, where they can quickly escalate

into major incidents. They are caused by releases of gaseous, flashing liquid and liquid inventories.

Jet fires develop rapidly, and the extreme heat they generate can result in structural failure if the flame impinges on critical members of the structure. This has important consequences for control and isolation strategies.

The properties of jet fires depend on:

▶ Fuel composition
▶ Release conditions
▶ Release rate
▶ Release geometry
▶ Wind direction and ambient conditions

Industry practice in assessment of jet fire hazards

Assessment of the hazards of jet fires is made by analysing the length of the jet flame relative to the distances of equipment, buildings, people, etc. Consideration is given to the extent to which the affected area is impinged on, as well as the necessity for Passive Fire Protection (PFP) and emergency depressurization, as well as other options in order to mitigate the hazard.

3.6.4 Flash fires and fireballs

When a dense cloud of vapour is formed by the release of flammable gases or liquids and this meets a source of ignition, a Vapour Cloud Fire (VCF) may result. This is also known as a flash fire or fireball.

Figure 3.7 Fire ball
Source: iStock.

Vapour Cloud Fires (VCFs) are important for two reasons:

▶ There is the possibility that they may escalate and cause secondary fires elsewhere.
▶ It is highly probable that a steady fire will follow a VCF, i.e. a pool or jet fire or a combination of both.

Flash fires and fireball modelling

The duration, height and diameter, as well as the amount of uplift, are all characteristics of flash fires and fireballs which can be modelled using a formula based on the mass of fuel released.

3.6.5 Explosion hazard assessment

On any installation there is a possibility of a wide variety of explosions, including:

▶ Unconfined explosions (overpressure generated by the presence of obstacles)
▶ Confined explosions (overpressure generated by a combination of confinement and obstacles)
▶ External explosions (associated with confined, vented explosions)
▶ Internal explosions (e.g. within a flare stack)
▶ Physical explosions (e.g. a failing pressure vessel)
▶ Solid phase explosions (e.g. those which are associated with the use of well completion explosives)
▶ Mist explosions
▶ Boiling Liquid Expanding Vapour Explosion (BLEVE)

On most installations, explosions are a significant component in the risk of topside fires. Data show that during the period 1992–1999 there were ten incidents involving explosions on UK installations. Most of the internal explosions were associated with flare systems or gas turbines.

3.6.6 Explosion hazard assessment modelling

Empirical models (knowledge acquired by means of observation or experimentation) employ a simplified version of the physics of an explosion and cannot deal with complex geometries. Consequently, their range of applicability is limited. However, they are useful for quickly calculating the order of magnitude of explosions, as well as for screening scenarios which then require further investigation with more sophisticated tools.

Phenomenological models (study of structures) are more complicated than empirical models. They are models which 'fit' the experimental data and are used to represent the scenario geometry in simple terms, e.g. boxes connected by corridors. They also have a limited range of applicability, less so than empirical models, although they provide a lower level of uncertainty.

Computational Fluid Dynamics (CFD) models fall into two groups – simple models and advanced models. Advanced models provide a more complete description of the physical and chemical processes involved in explosions, including an improved representation of the geometry and accuracy of the numerical schemes.

3.6.7 Explosion consequence assessment modelling

This type of modelling looks at the consequences of explosions. For example, injury to people from an explosion can be as a direct consequence of the blast wave (e.g. rupture of the ear drum) or indirectly (e.g. from flying debris). Equipment and structures can be damaged by the effects of loading (affecting walls and large objects) or drag loading (affecting steelwork or pipework which is constructed of narrow cross-sections) or a combination of both.

The extent of damage depends not only on the peak overpressure but also on the duration, impulse and rise time of the blast.

REVISION QUESTIONS FOR ELEMENT 1 CONTINUED

7 **Identify** the **FIVE** steps involved in undertaking a risk assessment.
8 **Explain** the concept of ALARP.
9 (a) **Explain** what a HAZOP is.
   (b) **Identify** who might be included in a HAZOP team and **give** the reason why they should be included.
10 There are a number of principles involved in inherent safe and risk based design concepts. **Give FOUR** examples.

You can compare your answers with our guidance at the end of the book under the section entitled 'Revision and assessment guide'.

Sub-element 1.4:

An organization's documented evidence to provide a convincing and valid argument that a system is adequately safe

This chapter covers the following learning outcome

Explain the purpose and content of an organization's documented evidence to provide a convincing and valid argument that a system is adequately safe in the oil and gas industries.

Chapter contents

4.1 Where documented evidence is used

4.1.1 Safety case and safety reports: an introduction

Organizations are obliged to produce documented evidence that their systems of operation, at any of their installations or facilities, are adequately safe. This document is known as a safety case or safety report. In some countries this is a legal requirement, but whether it is a legal requirement or not, it is still regarded as best practice and essential.

For offshore installations, there are regulations in the UK which are specific to the offshore oil and gas processing industry. These are the Offshore Installations (Safety Case) Regulations (OSCR) 2005. These regulations set out specific requirements to provide evidence and information that present a clear, comprehensive and defensible argument that a system is adequately safe to operate in a particular context. The document produced to provide this evidence is called a 'safety case'.

For onshore installations, there are regulations in the UK which are specific to the onshore oil and gas processing industry. These are the Control of Major Accident Hazards Regulations 1999 (COMAH) and their amendments 2005. Again, the regulations set out specific requirements to provide evidence and information that present a clear, comprehensive and defensible argument that a system is adequately safe to operate in a particular context. The document produced to provide this evidence is called a 'safety report'.

Similar legislation exists in some other countries around the world. However, for those countries that don't have specific legislation, similar standards exist because of the globalized nature of the oil and gas industry.

Regardless of which part of the world production activity is being undertaken, or what regulations apply, the requirements are very similar. Those requirements are that an organization should have in place a documented body of evidence to provide a convincing argument that a system is adequately safe to operate.

This is achieved by the provision and submission to the regulatory authority, where required, of a safety case or safety report.

4.2 The purpose of documented evidence such as safety cases and safety reports

Organizations are required to submit a safety case or safety report for each installation or facility they own or operate. This takes the form of a report which demonstrates the level of safety applied to that installation or facility. It has to be updated as required to reflect any changes in operational conditions.

The safety case or safety report covers all aspects of health and safety on an installation/facility. It is submitted at the planning stage and remains in place throughout the lifespan of the installation/facility until it is decommissioned. It is reviewed at five-yearly intervals by the enforcing authority, or sooner if requested. It should also be reviewed if there are significant modifications to the operation of the installation/facility.

4.3 The typical content of documents such as safety cases and safety reports

Let's look at offshore requirements first, and what is involved in writing a 'safety case'.

4.3.1 Offshore safety case

The safety case will be assessed by the enforcing authorities and evaluated in line with certain principles. These principles also reflect the structure and content of the safety case.

The three overriding principles to be demonstrated in a safety case are:

1 The management system is adequate to ensure compliance with statutory health and safety requirements; and for management of arrangements with contractors and sub-contractors.
2 Adequate arrangements have been made for audit and for audit reporting.
3 All hazards with the potential to cause a major accident have been identified, their risks evaluated, and measures have been, or will be, taken to control those risks to ensure that the relevant statutory provisions will be complied with.

The underlying principles are as follows:

Factual information

The safety case will include factual information about the installation itself, the plant and systems used, its location and external environment. It should also cover the activities to be carried out on, or in connection with, the installation. Each piece of information will be linked to all identified hazards associated with the information and which have the potential to cause a major accident.

Management of health and safety

The safety case should show how the management system will apply appropriate levels of control during each phase of the installation's life cycle. This will include the design, construction, commissioning, operation, decommissioning and dismantlement stages.

Control of major accident hazards

The safety case should demonstrate that all hazards with the potential to cause a major accident have been, or will be, identified, their risks evaluated and that measures have been, or will be, taken to control those risks.

Major accident hazard identification

The safety case will show how a systematic process has been, or will be, used to identify all reasonably foreseeable major accident hazards that are applicable to the installation. This will include identifying the initiating events or sequences of events related to those identified hazards.

Major accident risk evaluation

The safety case will clearly show what criteria have been, or will be, adopted for major accident risk assessment, including the methods used and the evaluation process applied. These will include:

▷ That particular attention has been, or will be, paid to instances or areas that have been identified where people may be exposed to significantly higher risks in comparison to the installation as a whole.

▷ How the evaluation has, or will, consider people as both a key element in safe operation as well as a potential cause of major accidents and their escalation.

▷ That adequate consideration of uncertainty has been, or will be, taken into account when presenting quantitative and qualitative risk assessment arguments.

▷ That the relative merits of engineering judgement and good practice have been, or will be, adequately considered.

▷ That the process of identifying risk reduction measures is systematic and takes into account new knowledge. Furthermore, what the reasoning was behind the choice of risk reduction measures.

▷ That proposed measures to reduce risk have timescales applied.

Major accident risk management

The safety case should describe what measures will be taken to manage major accident hazards. These will include:

▷ An explanation as to how inherently safer design concepts have been, or will be, applied in the decision-making process relating to design.

▷ What measures are in place to prevent major accident hazards during the installation's current phase of operation and the activities associated with it.

▷ What measures are provided for detecting events that require an emergency response.

▷ What control and mitigation measures will be provided to protect personnel from the consequences of a major accident. Also, how they will take account of likely conditions during an emergency. Finally, what measures and arrangements have been made for managing an emergency.

▷ What arrangements have been made to ensure that the Temporary Refuge (TR) will provide sufficient protection to enable people to muster safely.

▷ What arrangements and provisions have been made to ensure that the integrity of the Temporary Refuge (TR) is not compromised by any of the hazards identified in the risk assessment. Also, how long this integrity has been designed to be maintained for.

▷ Demonstrate that the evacuation and escape arrangements have been integrated in a logical and systematic manner. Also, that they take into account the potential worst environmental conditions in which they may need to be undertaken.

Rescue and recovery

The safety case should demonstrate that effective rescue and recovery arrangements have been planned for to cope with major accidents.

Life cycle requirements

The safety case should include a design notification, which describes how the principles of risk evaluation and risk management are being applied to the design to ensure that major accident risks will be controlled. This should include well engineering aspects, especially those that refer to well operations before the start of facility operations.

Combined operations

The safety case should demonstrate how the management system addresses the additional risks associated with combined operations.

These will include:

▷ Demonstration of a systematic approach to assessing the impact of combined operations on the conclusions of the operational safety case for each installation.

▷ Demonstration of a systematic approach to identify and assess any additional major accident hazards arising from combined operations.

Decommissioning and dismantlement

When the installation is reaching the end of its working life, the safety case will have to be revised to deal with decommissioning or dismantlement operations. At that point the safety case revision will include a description of the sequence of events from cessation of production to dismantling of the structure. The safety case will also include a description of the extent and availability of safety systems during decommissioning or dismantlement operations.

Any major accident hazards identified from the decommissioning or dismantlement operations will be identified in the safety case, as well as how the management system will maintain effective control during these periods.

4.3.2 Availability

The document should be made available to anyone on the installation.

4.3.3 Safe design concept

As we've mentioned, the report is required to include an explanation of how inherently safe design concepts were considered and applied. This requirement not only applies to when the installation was at the design stage but also at other stages in the life of the installation.

In order to ensure an inherently safer design, the design process should incorporate at a very early stage a hazard management strategy. This should include consideration of:

- The concept selection, e.g.
  - a platform or subsea development
  - attended or unattended wells
  - floating or fixed wells
  - multiple or single structures
  - the pre-drilling of wells
- Where the installation should be located and its orientation
- The substitution of hazardous substances and processes by less hazardous ones
- The segregation of hazards
- Reducing the complexity of the design
- The reduction of subsea uncertainty (e.g. the use of seismic surveys)
- The location and routing of the riser
- Making allowances for human factors (e.g. by designing in fail-safe features)
- The selection of materials
- The corrosion, erosion and stress concentration in the design
- How the design can allow for inspection and maintenance

4.3.4 Safety Management System (SMS)

The Safety Management System (SMS) is at the heart of the safety case. Once the aspects previously mentioned are implemented, it can be demonstrated that the management of safety incorporates a risk based approach, and that this is the basis of a safety management system.

The safety case should show that the safety management system is compliant with relevant statutory provisions.

The safety case should include the following elements in the descriptions of the management system in order to demonstrate that the system is adequate:

- Policy setting
  - outlining the policy and its objectives
  - demonstrating corporate acceptance of responsibility
- Organization
  - the structure of the organization
  - demonstration of its accountability
  - demonstration of its safety culture
  - demonstration of how professional health and safety advice will be shared
  - demonstration of how the workforce will be encouraged to be involved
  - outlining of the risk assessment systems
- Planning and standards
  - outlining the standards and procedures for controlling risks, including workload and working hours
  - outlining the permit-to-work system and where it will be applied
  - outlining how competency and training will be implemented
  - outlining how key personnel will be selected
  - outlining how control of change will be implemented
  - outlining how contractors will be selected and controlled
  - outline the planning and control for emergencies
  - outlining how occupational health will be managed
- Performance measurement
  - outlining how the recording and investigation of incidents will be implemented
  - outlining how active monitoring will be implemented
- Audit and review
  - outlining the auditing process
  - outlining when and how any review will be applied and the process for learning lessons

The safety management system should demonstrate an appropriate level of control during each phase of the installation's life cycle. This will include the design, construction, commissioning and operation as well as the decommissioning and dismantlement phases.

The safety management system should clarify who is in charge of activities during normal operating conditions and in emergency situations. This would include the arrangements for communications between the 'responsible persons' both on and offshore.

The safety management system should take account of:

- The levels of authority
- Performance standards

How to deal with exceptional conditions

Any lessons learned from previous incidents

In the situation where an installation is working in combination with another installation or vessel, the safety case should summarize any arrangements which have been put in place to co-ordinate both parties with the safety management system of the installation.

Let's now look at onshore requirements and what is involved in writing a 'safety report'.

4.3.5 Onshore safety report

The safety report will be assessed by the enforcing authorities and evaluated to ensure it meets the requirements of the legislation.

The safety report is split into five main sections:

1 Descriptive information
2 Information on management measures to prevent major incidents
3 Information on potential major incidents
4 Information on measures to prevent or mitigate the consequences of a major incident
5 Information on the emergency response measures of a major incident

Its content will be similar to that set out below:

Section 1 – descriptive information

An overview of the facility and its activities

The overview will give a general outline of the installation itself, what activities are carried out, and what products it uses and produces. It will also include the identified major incident scenarios and the measures in place for protection and intervention.

Information about dangerous substances in use at the facility

This will show:

The maximum quantities of dangerous substances likely to be present on the site at any time.

The chemical name of each and every type of dangerous substances involved in the process system.

The physical and chemical behaviour and/or characteristics of each type of dangerous substance including, where relevant, flashpoint, flammable limits, vapour pressure, density, etc.

The potential harm, either immediate or delayed, which could be caused by these dangerous substances, e.g. an asphyxiant, flammable, harmful to the environment, etc.

Information about the surrounding environment

A description of the surrounding environment including use of the land or activities conducted in the surrounding land, the extent and location of population, the location of significant buildings and infrastructure (e.g. hospitals, schools, road networks, etc.) and water extraction points.

A map of the area usually forms part of this section. This will also show the extent of the area to be affected by the worst case scenario.

Section 2 – information on management measures to prevent major incidents

Major Accident Prevention Policy (MAPP)

The Major Accident Prevention Policy (MAPP) sets out the policy on the prevention of major accidents and it should outline the following:

Description of the Safety Management System (SMS)

Roles and responsibilities of all key personnel

Training requirements to maintain competency levels as well as making good any identified shortfalls in competency levels

Hazard identification and risk assessment process

Procedures and instructions for the safe operation of plant

Design and any subsequent modification of the site

Identification of all foreseeable emergency scenarios and preparation for them

Accident investigation procedures

How compliance will be measured

Review and audit frequency and procedures

When, and under what circumstances, the Major Accident Prevention Policy (MAPP) is required to be updated

Section 3 – information on potential major incidents

This section describes the processes and scenarios that could lead to a major incident occurring. This will include details about the processes, the areas of the facility likely to be affected and the scenarios identified as plausible.

Section 4 – information on measures to prevent or mitigate the consequences of a major incident

This section describes the facility, the plant and the equipment in context of how major incidents can be prevented or mitigated. This will include details on operating parameters and what measures are in place to ensure they are not exceeded, emergency shutdown elements, detection equipment, fire-fighting arrangements, emergency evacuation and temporary refuge arrangements, etc.

All these elements will be categorized into either:

Inherent safety measures

Prevention measures

▷ Control measures
▷ Limitation measures

This section then expands to describe how safety and reliability have been built into the facility.

Section 5 – information on the emergency response measures of a major incident

Onsite emergency plan

This section describes the protection and intervention measures which are included in the onsite emergency plan. This will include

▷ What equipment there is to limit the consequences of a major incident
▷ What arrangements there are for alerting and intervening in an emergency
▷ What onsite and offsite resources are available
▷ What arrangements have been made to ensure all the resources and other equipment are maintained to an acceptable standard
▷ What arrangements there are for the training of personnel in emergency response
▷ What arrangements there are for testing the emergency plans

Offsite emergency plan

This section describes the arrangements for involving external emergency services and agencies. This will include:

▷ Details of the site including its location, roads and access points
▷ A site plan showing key facilities such as control centres, medical centres, main process plants and storage areas
▷ Details of site personnel
▷ Details of offsite areas likely to be affected by a major incident as well as levels of harm/damage possible. This will include types of buildings, population density, sensitive buildings, drainage detail, etc.
▷ Details of dangerous substances on site including types of substances, quantities, hazardous properties, location, etc.
▷ Details of any relevant technical advice
▷ Details of equipment and resources that are available for fire-fighting purposes
▷ The function of key posts with duties in an emergency response, their location and how they can be identified
▷ An outline of the initial actions to be taken in case of an emergency situation, such as warning the public, setting up emergency facilities such as a control room, etc.

Sources of reference for Element 1

Atherton, J. and Gil, F. for Centre for Chemical Process Safety (CCPS) and American Institute of Chemical Engineers (2008) *Incidents That Define Process Safety*, New Jersey: John Wiley & Sons. ISBN 978-0-470-12204-4.

BP U.S. Refineries Independent Safety Review Panel (2007) *Report of the BP U.S. Refineries Independent Safety Review Panel*. Available at: www.bp.com/liveassets/bp_internet/globalbp/globalbp_uk_english/SP/STAGING/local_assets/assets/pdfs/Baker_panel_report.pdf

British Standards Institution (2001) *ISO 17776 Petroleum and Natural Gas Industries. Offshore Production Installations. Guidance on Tools and Techniques for Hazard Identification and Risk Assessment*. London: BSI. ISBN 0 580 36915 3.

Buncefield Major Incident Investigation Board (2013) *The Buncefield Incident 11 December 2005. The Final Report of the Major Investigation Board Volume 2*. London: HSE Books. ISBN 978-0-7176-6318-7. Available at: www.buncefieldinvestigation.gov.uk/reports/volume2a.pdf

Cullen, The Honourable Lord, W. D. (1990) *The Public Inquiry into the Piper Alpha Disaster* (Command Paper). London: Department of Energy. ISBN 978-0-101-13102-5.

Dawson, D. and Brooks, B. (1999) *Report of the Longford Royal Commission: The Esso Longford Gas Plant Accident*. Melbourne: Government Printer for the State of Victoria.

Energy Institute (2008) *Guidance on Investigating and Analysing Human and Organisational Factors Aspects of Incidents and Accidents*. London: Energy Institute. ISBN 978-085293-521-7. Available at: http://www2.energyinstpubs.org.uk/pdfs/817.pdf

Gadd, S., Keeley, D. and Balmforth, H. (2003) *Good Practice and Pitfalls in Risk Assessment*. London: HSE Books. ISBN 0-7176-2732-2. Available at: www.hse.gov.uk/research/rrpdf/rr151.pdf

Health and Safety Executive (n.d.) 'ALARP at a Glance'. HSE Guidance. Available at: www.hse.gov.uk/risk/theory/alarpglance.htm

Health and Safety Executive (1999) *Reducing Error and Influencing Behaviour* (HSG48). 2nd edn. London: HSE Books. ISBN 978-0-7176-2452-2. Available at: www.hse.gov.uk/pubns/priced/hsg48.pdf

Health and Safety Executive (2005) *Preparing Safety Reports: Control of Major Accident Hazards Regulations 1999* (HSG190). London: HSE Books. ISBN 978-0-7176-1687-9.

Health and Safety Executive (2006a) *A Guide to the Offshore Installations (Safety Case) Regulations 2005*. London: HSE. ISBN 978-0-7176-6184-8 (L30). Available at: www.hse.gov.uk/pubns/books/l30.htm

Health and Safety Executive (2006b) 'Assessment Principles for Offshore Cases (APOSC)'. London: HSE. Available at: www.hse.gov.uk/offshore/aposc190306.pdf

Health and Safety Executive (2008) 'Modelling of Pool Fires in Offshore Hazard Assessments Offshore – Information Sheet No. 9/2008'. London: HSE. Available at: www.hse.gov.uk/offshore/infosheets/is9-2008.pdf

Health and Safety Executive (2009) *Safety and Environmental Standards for Fuel Storage Sites: The Process Safety Leadership Group Final Report*. London: HSE. ISBN 978-0-7176-6386-6. Available at: www.hse.gov.uk/comah/buncefield/fuel-storage-sites.pdf

Health and Safety Executive (2011) *Investigating Accidents and Incidents* (HSG245). 2nd edn. London: HSE Books. ISBN 978-0-7176-2827-8. Available at: www.hse.gov.uk/pubns/priced/hsg245.pdf

Kletz, T. A. (1998) *What Went Wrong? Case Histories of Process Plant Disasters*. Oxford: Gulf. ISBN 0-88415-920-5.

Kletz, T. A. (2001) *Learning from Accidents*. 3rd edn. Oxford: Gulf. ISBN 0-7506-4883-X.

Kletz, T. A. (2003) *Still Going Wrong: Case Histories of Process Plant Disasters and How They Could Have Been Avoided*. Oxford: Gulf. ISBN 0-75067-709-0.

Mansfield, D., Poulter, L. and Kletz, T. (1996) *Improving Inherent Safety* (OTH 96 521). London: HSE Books. ISBN 0-7176-1307-0. Available at: www.hse.gov.uk/research/othpdf/500-599/oth521.pdf

New South Wales Department of Planning (2011) 'Hazardous Industry Planning Advisory Paper No. 8: HAZOP Guidelines'. Sydney: State of New South Wales. ISBN 978-0-73475-872-9. Available at: www.planning.nsw.gov.au/Portals/0/HIPAP%208%20Final%202011.pdf

U.S. Chemical Safety and Hazard Investigation Board (2007) *Investigation Report: Refinery Explosion and Fire, BP Texas City, Texas*. Report No. 2005-04-1-TX. Available at: www.csb.gov/assets/1/19/csbfinalreportbp.pdf

Veritas, D. N. for the Health and Safety Executive (2002) *Marine Risk Assessment*. London: HSE Books. ISBN 0-7176-2231-2. Available at: www.hse.gov.uk/research/otopdf/2001/oto01063.pdf

4

Sub-element 2.1:
Contract management

5.1 Scale of contractor use

Contractors are used extensively in the oil and gas industry for all kinds of work and at every stage of the operation. This extends to, and includes, exploration, construction, production and decommissioning, as well as the less technical services such as cleaning, catering and waste removal. They can also be used for emergency response.

Contractors tend to be used for specialist work, or work that is specific to a time when there is a raised workload such as maintenance during shutdown, inspection or testing.

For example, when an installation has to be shut down within a tight schedule, the number of extra staff required on site at any particular time may increase to as many as 200. Consequently, the exposure to hazards will increase due to the higher number of personnel on the installation.

Depending upon the requirements for contractors, they can vary from one-man operators to multi-national companies employing large numbers of staff. Some of the contractors may be ex-employees who are employed on a contractual basis during shutdowns – they have the advantage of having experience and knowledge of the installation and its operating procedures (e.g. mechanical technicians).

5.2 Contractor management, ownership and site supervision/representation

It is essential that the client (employing company) and the contractor both have a clear understanding of their roles and responsibilities in order for a project to be carried out safely and successfully. These are outlined below.

5.2.1 Client (employing company)

The client is responsible for the workplace the contractor is working in

Clients are those organizations who select and engage contractors to undertake specific work for them. They are obliged to protect their own employees, as well as the employees of the contractors they are engaging, from risks to health and wellbeing. This will probably be underpinned by local legislation.

The client also has a responsibility towards the contractor and the contractor's own staff for hazards that may occur as a result of the client's own activities.

An example might be where a refining company loads its vehicles in separate bays and has a need to have those bays checked to see if their electrics are safe and up to standard. It decides to engage a specialist electrical company to test the equipment.

A safe method of working is then agreed between the client and the contractor which allows the contractor to work in one loading bay at a time. This then allows the other bays to remain operational whilst the checks are carried out in one bay.

The movement of vehicles within the general area of the loading bays will pose the biggest hazard to the contractor. The client will need to carry out an induction with the staff of the electrical company who will be working in the bays. This would cover the movement of vehicles, places that the contractor should and should not have access to, etc.

5.2.2 Contractors

The contractor is responsible for the safe method of working

Contractors are those people or organizations who have been selected and engaged to undertake work for the client.

The contractor will not be as familiar with the workplace and its hazards as regular workers. Consequently, adequate control measures should be implemented in order to counteract this increased risk. Examples might be:

▷ Limiting access the contractor has to specific areas
▷ Giving all of the contractor's staff a detailed induction course on specific hazards likely to be encountered

5.2.3 Procedures for selecting suitable contractors

Clients need to ensure that any contractor they engage is competent. That is to say, they have sufficient skills and knowledge to undertake and complete the work safely, efficiently and to the quality level required. The degree of competency required will inevitably depend on the work to be done.

The client should ensure contractors are fully aware of what is expected of them. Health and safety arrangements should be explained to them. Contractors should also be instructed on standard procedures, any permit-to-work systems, and the current health and safety policy statement.

Determining a contractor's competence should be the first step in choosing a suitable contractor. The kinds of issues which should be considered include the following:

▷ Is the contractor adequately insured?
▷ Has, or will, the contractor undertake a risk assessment of the proposed contracted work?
▷ Are the health and safety policies and practices of the contractor adequate?
▷ Is the contractor's recent health and safety performance reasonable (number of accidents, etc.)?
▷ Is the contractor's health and safety training and supervision adequate?
▷ Does the contractor have arrangements in place for consulting with their workforce?
▷ Does the contractor or their individual employees hold a 'passport' or other type of certification in health and safety training?
▷ What, if any, enforcement notices have been served on the contractor?
▷ Can the contractor offer any independent assessment of their competence?
▷ What references from previous employing companies can the contractor show?
▷ What relevant qualifications and skills does the contractor have?
▷ What is the level of competency of the staff doing the job?
▷ Is there appropriate certification for any equipment that the contractor might intend to use?
▷ What selection procedure does the contractor have for sub-contractors they might engage with?
▷ Is the contractor a member of a relevant trade or professional body?
▷ What is the contractor's financial viability?
▷ Does the contractor's safety method statement meet expectations?

5.2.4 Managing/supervising contractors

Those who are responsible for managing contractors must ensure their relationship with contractors is handled appropriately. As we have just discussed, rigorous checks must be made to ensure the contractor is suitable for the work being undertaken. Once a contractor is selected and engaged, the client should provide an appropriate level of supervision of the work being undertaken by the contractor.

The level of supervision will depend on the size and/or complexity of contracted work. For example, maintenance work may require constant supervision to ensure the agreed safe working procedures are being adhered to.

Larger contracts, on the other hand, may need a more structured means of management control. This might involve the contractor relaying daily reports on all aspects of the operation.

It is for the clients to decide what controls they need to implement in order to effectively manage and supervise the work of the contractor. The more impact the contractor's work potentially could have on the health and safety of anyone likely to be affected, the greater the management and supervisory responsibilities of the client.

Clients also carry a greater responsibility when their knowledge about the health and safety implications of the contracted work is more extensive than that of the contractor.

It is essential that the nature and extent of the controls exercised by the client in managing the contracted work is agreed by both parties before work starts. Clients may need to agree with the contractor on how the work will be done and the precautions that will be taken. This might include the following:

- What equipment should or should not be worked on/used
- Personal protective equipment to be used and who will provide it
- Working procedures, including any permit-to-work systems
- Training and induction in such systems
- The number of people needed to do the job
- The reporting of accidents and safe keeping of records and plans

Clients, contractors and sub-contractors should monitor their health and safety performance. This will include ensuring risk assessments are current, and that control measures are being applied and are effective. The degree of risk involved with the contracted work will govern the level of monitoring required.

Let's take a look at a couple of case studies that relate to contract management.

Case Study 5.1

A client engaged a contractor to do some maintenance work on an electrical substation. However, the client failed to check the competency of the contractor to undertake the work. This was compounded when the client failed to supervise the work adequately and to control access to the substation.

Within the substation, open fuse-boards were inadequately shrouded. The result was that one of the contractor's employees was severely injured when he came into contact with a live piece of equipment.

When the enforcing authorities prosecuted the client and the contractor, the client was found guilty of not applying due diligence in selecting or supervising the contractor and was fined £50,000 plus costs.

The contractor was found guilty of failing to adequately control the installation of two ceiling fans in the substations and was fined £30,000 plus costs.

The experienced electrician employed by the contractor and who was supervising the activities of the electrician who was injured was found guilty of failing to take simple steps to prevent contact with live equipment and was fined £1,000 plus costs.

Case Study 5.2

An onshore facility needed to replace some bolts on a pipe flange which had rusted quite badly. The pipework involved was used to convey high temperature chemicals and, as the pipe could not easily be shut down, it was agreed that the safe working procedure should be to remove one bolt at a time and to replace each bolt before removing and replacing the next bolt.

The company contracted an ex-employee to supervise the work and he went with the fitter to the site and they had a toolbox talk covering the safe working procedure previously agreed. The contractor then left the fitter to get on with the job.

Once the fitter started the work it soon became clear the bolts would not undo using a spanner as they were too rusty. He decided he needed to use a hacksaw. He also decided to remove all of the old bolts before replacing them.

When the last bolt was removed, the flange fell off and the fitter was sprayed with hot chemical liquid, causing severe burns.

The contractor had been employed purely to supervise this work but by only giving a toolbox talk prior to the work being started, the level of supervision was far from adequate.

5.3 Contractor responsibilities

As we mentioned in the previous section, it is **the contractor who is responsible for the safe method of working he/she and his/her staff employ**.

This is because, in many cases, they are being engaged because of their specialist knowledge and expertise, whereas the company contracting out the work does not have that expertise.

For example, maintenance work needed on an offshore installation's subsea structure would need specialist divers to undertake the work. This would be beyond the experience of the normal workforce on the installation.

Contractors also need to give an assurance that the work they carry out will not impact on the safety of operations (e.g. will not shut down emergency systems such as alarms, emergency shutdown systems, etc.).

5.3.1 Other responsibilities of the contractor

The contractor also has other responsibilities, particularly where a contractor employs sub-contractors. In this case, both the contractor and sub-contractor will have some health and safety responsibilities. The extent of the responsibilities of each party will depend upon the circumstances.

Although the selection and employment of sub-contractors is ideally the role of the contractor, it is ultimately the responsibility of the client to ensure that the contractor is qualified and has an effective procedure in place with which he/she can assess the competence of a sub-contractor. The contractor may use similar criteria to select a sub-contractor to that used by the client when selecting a main contractor. The work to be done by the sub-contractor will determine the level of competence required.

Other responsibilities of the contractor: risk assessment

There is a requirement for a risk assessment to be conducted concerning the work which is to be undertaken by the contractor. There also needs to be agreement about the measures that will be applied to control those risks identified in the risk assessment. These need to be agreed between the client and the contractor as well as any sub-contractors. Consideration should also be given to any risks which could arise from each other's work that could impact on the health and safety of employees or any other persons affected by that work.

Other responsibilities of the contractor: instruction and training

There is a requirement for contractors and sub-contractors (as well as the client) to:

▷ Consider what information should be passed between them and agree appropriate ways to ensure this is done.

▷ Exchange clear information about any risks arising from their work, including relevant safety rules and procedures. These may include permit-to-work systems and arrangements for dealing with emergencies.

▷ Provide any information, instruction and training to their employees on anything which may impact on their health and safety. This may include induction or training on permit-to-work systems.

▷ Consult their employees (and safety representatives if applicable) on health and safety matters.

Contractors and sub-contractors should monitor their health and safety performance. This allows them to check if their risk assessments are up to date and whether existing control measures are working or if further control measures need to be implemented. The level of monitoring needed is relative to the level of risk, i.e. the higher the risk, the greater the level of monitoring that will be required.

5.4 Safe handover: understanding the hazards

5.4.1 Safe handover procedure

Whenever a site, or part of a site, piece of machinery or plant, is to be handed over to a contractor for them to undertake their work, a comprehensive system of safety checks should be implemented. This may include:

▷ Electrical isolations with lock offs as appropriate

▷ Mechanical isolations with lock offs as appropriate

▷ Physical barriers to restrict access by non-authorized personnel

▷ Pre-cleaning of the area

▷ Pre-cleaning of the equipment

Whatever system is implemented, it needs to be agreed between both the client and the contractor. It must also take into account the arrangements and procedures of both parties. However, as the work is likely to be undertaken on the client's premises or installation, this may mean that the client's arrangements and procedures are used as a basis for the agreed procedure, with the contractor's specific requirements superimposed on those procedures.

There should be a record kept of what was included in the handover. Particularly important are any isolations and barriers which may be involved.

Also to be recorded is the acceptance by the contractor of responsibility to maintain the integrity of the area during the period of the contract. This should be part of the handover process. There is a need, however, for both parties to carry out periodic checks during that period to make sure that any safety measures which have been put in place are complied with.

Once the contract is completed, there should be a procedure for handing the site, or part of a site, piece of machinery or plant, back to the client. This handback should follow an agreed procedure to ensure that all matters have been dealt with safely and securely. This will include confirming that all parts of the installation which have been temporarily the responsibility of the contractor are in full working order and also that any isolations and barriers have been removed. Any pipework which has been worked on must be pressure tested to confirm its integrity (usually by using either water or an inert gas).

Once all of the criteria for the handover process have been completed and the client is satisfied that everything is in working order, then the contract will be signed off and the plant re-commissioned. It is essential for both parties that the handback process is conducted thoroughly and meticulously.

REVISION QUESTIONS FOR ELEMENT 2

1 **Identify EIGHT** factors that should be considered when selecting a suitable sub-contractor.

You can compare your answers with our guidance at the end of the book under the section entitled 'Revision and assessment guide'.

5

Sub-element 2.2:
Process safety management

6.1 The controls available in process safety management

6.1.1 Process Safety Management (PSM)

Process Safety Management (PSM) is the management of hazards associated with the processing of products, particularly highly volatile substances such as hydrocarbons. The principle behind it is to reduce the number of incidents involving the release of highly volatile or toxic substances or, at the very least, mitigate the severity of such incidents.

The oil and gas industries use process safety management extensively, particularly where they are processing volatile products or have large inventories of flammable or toxic materials.

'Process' in this respect means using, storing, manufacturing, handling or moving a hazardous product/chemical on a site or installation.

Process safety management – what are the key benefits?

The main aim of Process Safety Management (PSM) is to develop plant systems and procedures which will prevent unwanted releases which may ignite and/or cause toxic impacts, local fires or explosions.

In addition, PSM can also address issues related to the operability, productivity, stability and quality output of processes.

If we look at the major incidents that have occurred over the past thirty years, such as Flixborough, Piper Alpha, Buncefield, Texas City, Deepwater Horizon, etc., we can understand the significance of applying PSM as a proactive tool.

The US Department of Labor Occupational Safety and Health Administration (OSHA) has published, and regularly updates, guidance on process safety management (OSHA 3133) which the oil and gas industry tends to use as a template for developing their own systems.

The key provision of PSM is Process Hazard Analysis (PHA). This is basically a careful review of what could go wrong and what safeguards must be implemented to prevent releases of hazardous substances.

Prior to conducting a process hazard analysis, management should complete a compilation of written process safety information. This compilation will help the management and the employees involved in operating the process to identify and understand the hazards posed by those processes.

Process safety information should include information on the hazards of the products used or produced by the process, information on the technology of the process and information on the equipment in the process. Let's now look at the extent of information needed under these categories.

Process safety management – preparation for a process hazard analysis

Information on the hazards of the products/chemicals in the process should consist of the following:

▶ Toxicity
▶ Permissible exposure limits
▶ Physical data
▶ Reactivity data
▶ Corrosivity data
▶ Thermal and chemical stability data, and hazardous effects of the inadvertent mixing of different materials

Information on the technology of the process must include at least the following:

▶ A process flow diagram
▶ Process chemistry
▶ Maximum intended inventory
▶ Safe upper and lower limits for such items as temperatures, pressures, flows or compositions
▶ An evaluation of the consequences of deviations, including those affecting the safety and health of employees

Information on the equipment in the process should include the following:

▶ Materials of construction
▶ Piping and instrument diagrams (P&IDs)
▶ Electrical classification
▶ Relief system design and design basis
▶ Ventilation system design
▶ Design codes and standards employed
▶ Safety systems (e.g. interlocks, detection or suppression systems)

The compilation of the above process safety information provides the basis for identifying and understanding the hazards of a process and is necessary in developing the Process Hazard Analysis (PHA). It may also be necessary for other provisions of process safety management such as management of change and incident investigations.

Process safety management – Process Hazard Analysis (PHA)

Process Hazard Analysis (PHA) is a thorough, orderly, systematic approach for identifying, evaluating and controlling the hazards of processes involving hazardous products/chemicals. Management should perform an initial process hazard analysis (hazard evaluation) on all processes. The PHA methodology selected must be appropriate to the complexity of the process and must identify, evaluate and conclude with appropriate controls relating to the hazards involved in the process.

Management should determine and document the order of priority for conducting process hazard analysis based on a rationale that includes such considerations as the extent of the process hazards, the number of potentially affected personnel, the age of the process and the operating history of the process.

Management should use one or more of the following methods, as appropriate, to determine and evaluate the hazards of the process being analysed:

▶ What-if/checklist
▶ Hazard and Operability Study (HAZOP)
▶ Hazard Identification Study (HAZID)
▶ Failure Mode and Effects Analysis (FMEA)
▶ Fault tree analysis
▶ Any other appropriate equivalent methodology

Whichever method(s) are used, the PHA must address the following:

▶ The hazards of the process
▶ The identification of any previous incident that had a potential for catastrophic consequences in the workplace
▶ Engineering and administrative controls applicable to the hazards and their interrelationships, such as appropriate application of detection methodologies to provide early warning of releases. Acceptable detection methods might include process monitoring and control instrumentation with alarms, and detection hardware such as hydrocarbon sensors.
▶ Consequences of failure of engineering and administrative controls
▶ The siting of the facility
▶ Human factors
▶ A qualitative evaluation of a range of the possible safety and health effects on employees in the workplace if there is a failure of controls

The process hazard analysis is best performed by a team with expertise in engineering and process operations, and the team should include at least one employee who has experience of, and knowledge of, the process being evaluated. Also, one member of the team should be knowledgeable in the specific analysis methods being used.

Management should:

▶ Establish a system to address promptly the team's findings and recommendations
▶ Ensure that the recommendations are resolved in a timely manner and that the resolutions are documented
▶ Document what actions are to be taken
▶ Develop a written schedule of when these actions are to be completed
▶ Complete actions as soon as possible
▶ Communicate the actions to operating, maintenance and other employees whose work assignments are in the process and who may be affected by the recommendations or actions

The process hazard analysis should be reviewed regularly and updated when necessary to ensure that the hazard analysis is consistent with the current process.

Process safety management – operating procedures

Management should develop and implement written operating procedures, consistent with the process safety information, that provide clear instructions for safely conducting activities involved in each covered process. The tasks and procedures related to the covered process must be appropriate, clear, consistent and, most importantly, well communicated to employees. The procedures should address the following elements:

Steps for each operating phase

▶ Initial start-up
▶ Normal operations
▶ Temporary operations
▶ Emergency shutdown, including the conditions under which emergency shutdown is required, and the assignment of shut down responsibility to qualified operators to ensure that emergency shutdown is executed in a safe and timely manner
▶ Emergency operations
▶ Normal shutdown
▶ Start-up following a turnaround, or after an emergency shutdown

Operating limits

▶ Consequences of deviation
▶ Steps required to correct, or avoid, deviation

Health and safety considerations

▶ Properties of, and hazards presented by, the products/chemicals used in the process

▶ Precautions necessary to prevent exposure, including engineering controls, administrative controls and personal protective equipment
▶ Control measures to be taken if physical contact or airborne exposure occurs
▶ Quality control for raw materials and control of hazardous chemical inventory levels
▶ Any special or unique hazards
▶ Safety systems (e.g. interlocks, detection or suppression systems) and their functions

Management should ensure up-to-date references are available to employees who work in or maintain a process. The operating procedures must be reviewed as often as necessary to ensure that they reflect current operating practices, including changes in process chemicals, technology, equipment and facilities.

Management should develop and implement safe work practices to provide for the control of hazards during work activities, such as safe working procedures, permit-to-work systems, lockout/tag-out, confined space entry, opening process equipment or piping and control over the entry into a facility by maintenance, contractor, laboratory or other support personnel. These safe work practices and procedures should apply both to employees and to contractors.

Process safety management – employee participation

Management should consult with their employees and/or their representatives on the conduct and development of process hazard analysis and on the development of the other elements of process management. They should also provide access to process hazard analysis and to all other relevant information.

Process safety management – training

For Process Safety Management (PSM) to be effective, it is necessary for all personnel involved in the process to be fully trained on the implementation of the safe working procedures which have evolved from the Process Hazard Analysis (PHA) procedure.

All employees involved in operating a process, or a newly assigned process, should be trained in an overview of the process and in its operating procedures. The training should include an emphasis on the specific hazards of the process, emergency operations including shutdown, and other safe work practices that apply to the employee's job or tasks.

Process safety management – contractors

It is essential that contractors and their employees emphasize the importance of everyone taking care that they do nothing to endanger those working nearby who may work for another employer.

6

Clients need to ensure that any contractor they engage is competent. That is to say, that they have sufficient skills and knowledge to undertake and complete the work safely, efficiently and to the quality level required.

Once a contractor is selected and engaged, the client should provide an appropriate level of supervision of the work being undertaken by the contractor. However, it is the contractor who is responsible for the safe method of working which he/she and his/her staff employ.

Process safety management – pre-start-up safety review

It is important that a safety review takes place before any hazardous product is introduced into a process. Consequently, management should perform a pre-start-up safety review for new facilities and for modified facilities when the modification is significant enough to require a change in the process safety information. Prior to the introduction of a hazardous product to a process, the pre-start-up safety review must confirm the following:

▶ Construction and equipment are in accordance with design specifications
▶ Safety, operating, maintenance and emergency procedures are in place and are adequate
▶ A process hazard analysis has been performed for new facilities and recommendations have been resolved or implemented before start-up, and modified facilities meet the management of change requirements
▶ Training of each employee involved in operating a process has been completed

Process safety management – mechanical integrity

It is important to maintain the mechanical integrity of critical process equipment to ensure it is designed and installed correctly and operates properly. Under process safety management techniques and systems, mechanical integrity applies to the following equipment:

▶ Pressure vessels and storage tanks
▶ Piping systems (including piping components such as valves)
▶ Relief and vent systems and devices
▶ Emergency shutdown systems
▶ Controls (including monitoring devices and sensors, alarms and interlocks)
▶ Pumps

Management should establish and implement written procedures to maintain the ongoing integrity of process equipment. Employees involved in maintaining the ongoing integrity of process equipment should be trained in an overview of that process and its hazards, and trained in the procedures applicable to their particular job or task.

Inspection and testing should be performed on process equipment, using procedures that follow recognized, and generally accepted, good engineering practices. The frequency of inspections and tests of process equipment should conform, at the very least, to the manufacturers' recommendations and good engineering practices. Each inspection and test on process equipment should be documented, identifying the date of the inspection or test, the name of the person who performed the inspection or test, the serial number or other identifier of the equipment on which the inspection or test was performed, a description of the inspection or test performed and the results of the inspection or test.

Equipment deficiencies beyond the acceptable limits defined by the process safety information should be corrected before further use or in a timely manner if appropriate.

When constructing new plant and equipment, management should ensure that equipment is suitable for the process application for which it will be used. Appropriate checks and inspections should be undertaken to ensure that equipment is installed properly and is consistent with design specifications and the manufacturers' instructions.

Management should also ensure that maintenance materials, spare parts and equipment are suitable for the process application for which they will be used.

Process safety management – permits-to-work

A permit-to-work must be issued prior to any work or operations being conducted on or near a process. This includes any hot work, breaking of containment, confined space entry, working at height or over water, etc.

Process safety management – management of change

Any change to a process must be thoroughly evaluated in order to fully assess its impact on employee health and safety, as well as to determine what adjustments will be needed to operating procedures. Management should ensure that the following considerations are addressed prior to any change being implemented:

▶ The technical basis for the proposed change
▶ Impact of the change on employee health and safety
▶ Modifications to operating procedures
▶ Necessary time period for the change
▶ Authorization requirements for the proposed change

Employees who operate a process, as well as maintenance and contract employees whose job or tasks will be affected by a change in the process, must be informed of, and trained in, the change prior to start-up of the process.

We will be looking in more detail at management of change controls later in this chapter.

Process safety management – incident investigation

Each and every incident that released, or could have resulted in a release of, any flammable or toxic substance should be investigated. A system must be established to promptly address and resolve the incident investigation's findings and recommendations.

Process safety management – emergency planning and response

If, despite the best planning, an incident does occur, it is essential that a predetermined emergency plan is put into action. Again, all employees, as well as contractors, should be aware of what the plan is and what their role in that plan is. We shall be looking at emergency response in more detail in Chapter 18.

Now we have covered the principles of process safety management, we need to see how these principles are put into practice.

In particular we are going to look at:

▶ **Plant layout**
▶ **Control room design**
▶ **Temporary refuge integrity**

6.1.2 Plant layout

Plant layout is often a compromise between various factors. These can include:

▶ The need to locate hazardous materials in facilities as far from site boundaries and local residents as possible. This may be limited by the geography of the site.
▶ The need to minimize distances between buildings and storage areas where materials are located in order to reduce the risks during transfer.
▶ The need to facilitate interaction between other plants on the same site, allowing for greater operability and maintainability.
▶ The need to provide access for emergency services as well as emergency escape routes for personnel working on the site. It is also necessary to prevent confinement in the case of accidental release of flammable substances.
▶ Interaction with existing facilities, e.g. roads, drainage and utilities.
▶ The need for acceptable working conditions for personnel.

With regard to safety, the most important factors of plant layout are those which:

▶ Prevent or mitigate an escalation of events
▶ Ensure safety of personnel within on site buildings
▶ Control access of unauthorized personnel
▶ Facilitate access of emergency services

When designing plant layout, consideration should be given to the following factors:

▶ Inherent safety
▶ Fire
▶ Explosion
▶ Toxic gas release
▶ How to reduce the consequences of an event
▶ Positioning of occupied buildings
▶ The aggregation or trapping of flammable vapour

Let's look at each of these in more detail.

Plant layout – inherent safety

The first principle in applying inherent safety to a plant layout is to remove the hazard altogether. This means reducing the inventory of hazardous substances within the plant layout to such a level that a major hazard is no longer exists.

However, this is not often achievable, as the sole purpose of many installations is to process hazardous substances in one way or another.

Other possible methods to achieve an inherently safe design include:

▶ Reducing inventories wherever possible.
▶ Substituting hazardous substances with less hazardous alternatives where possible.
▶ Minimizing hazardous process conditions where possible i.e. temperature, pressure, rate of flow, etc.
▶ Designing systems and processes to be as simple as possible. This will have the effect of reducing human errors which could create a hazardous event.
▶ Using fail-safe design features, e.g. a downhole safety valve which requires hydraulic pressure to open the valve and allow product to flow, but in the case of lack of hydraulic pressure, the valve closes and product stops flowing.

Plant layout – the Dow/Mond indices

These indices are useful in the development stage of a project because they evaluate process plant hazards and rank them against existing processes or projects in order to provide a comparative measure of the risk of fire and explosion. They do this by assigning them incident classifications and allow objective spacing distances to be considered during the development phase of a process or project.

Plant layout – domino effects

When undertaking a hazard assessment of site layout it is essential that the consequences of loss of containment are fully evaluated. This includes the potential escalation of an incident and what this entails. These escalations are known as a domino effect, an example of which may be a loss of containment followed by fire and explosion which in turn damages the control room causing the loss of control of operations in other locations.

Plant layout – fire

There are four ways a fire can spread:

▶ Direct burning
▶ Convection
▶ Radiation
▶ Conduction

Fire can be prevented from spreading by fire resistant walls, floors and ceilings.

However, running fires caused by flammable liquids can spread through drains, ducts and ventilation systems. Consideration of these possibilities should be given at the plant layout design stage. Furthermore, flammable gases and vapours may also find their way into passageways and cause a delayed ignition.

Other considerations to reduce the effects of convection, conduction and radiation during a fire include ensuring the distances between plant items are sufficient in order to prevent overheating of adjacent plants. Where this is not possible, other methods such as fire walls, active or passive fire protection may be considered.

Plant layout – explosion

An explosion can be described as the release of energy in a violent manner (pressure wave) accompanied by the generation of high temperatures. This release of energy can often create missiles which emanated from the source of the explosion.

With regard to explosion pressure waves when designing plant layout, the following are mitigating factors:

▶ Ensuring separation distances are sufficient in even the worst case so that damage to adjacent plants will not occur
▶ Providing blast walls or locating adjacent plant in strong buildings
▶ Ensuring the walls of vessels are thick and strong enough to withstand a pressure wave from an explosion
▶ Ensuring any explosion relief vents are directed away from vulnerable areas, e.g. other plants or buildings, or roadways near site boundaries

The consequences of missiles emanating from an explosion will require a more involved and detailed analysis on an individual basis in order to provide adequate safety.

Plant layout – toxic gas releases

The release of toxic gas may render a plant or process inoperable due to the domino effect which may ensue. This could be as a result of personnel being injured and unable to bring the situation under control.

This may be prevented or mitigated by:

▶ The implementation of automatic control systems (which use inherently safer principles)
▶ Controlling plant and processes from a remote facility, e.g. a suitably protected control centre

Plant layout – reduction of consequences of event on- and offsite

With regard to the design of inherently safe plant layout, other measures which may be considered include the following:

▶ Storage of flammable/toxic material away from process areas.
▶ Siting hazardous plant and processes away from main roadways within the site.
▶ Fitting remote-actuated isolation valves where large inventories of hazardous materials may be released.
▶ Using the terrain as a means of controlling potential releases of liquid hazardous material. This can include embankments (bunds), ditches, dykes, etc.
▶ The siting of plant within buildings to act as a secondary containment facility.
▶ Where there is the potential for minor release of flammable gases or vapours, then siting this plant in the open air. This will ensure there is a rapid dispersion of these gases or vapours.
▶ Classification of areas where flammable gases, vapours and dusts may be released. This will enable ignition sources to be controlled and eliminated.

Risk management techniques will enable the appropriate control measures to be applied.

Plant layout – positioning of occupied buildings

As we mentioned earlier, the distance between, and the position of, occupied buildings and buildings housing plant will be governed by the need to reduce the domino effect of a fire, an explosion or the release of toxic gases. Over and above the distance and proximity issues, consideration should also be given to evacuation routes, which should not be impeded by poor plant layout. The siting of occupied buildings should generally be upwind of hazardous plant areas.

Plant layout – aggregation/trapping of flammable vapours

In order to prevent flammable/toxic vapours accumulating in buildings and creating a hazardous event, buildings should be well ventilated by natural or engineered ventilation.

Storage of flammable products should be in the open air so that any minor leaks can be allowed to disperse naturally, although some form of warning system should be deployed so that a minor leak does not develop into a major leak.

6.1.3 Control room design

With regard to control room design, there are two main aspects that should be part of the design consideration. These are:

▶ The ability of the control room to withstand a major hazardous event such as a fire, explosion or release of toxic gas or smoke.

▶ The efficient and appropriate layout of the control room and its equipment to ensure the effective operation and control of the plant under any circumstances, including an emergency.

Control room design – control room structure

The structure of the control room should be such that it can be safely used by personnel to maintain plant control at all times, even if an emergency situation or undesirable event takes place.

Events that have the potential to affect the control room are:

▶ Vapour Cloud Explosion (VCE)
▶ Boiling Liquid Expanding Vapour Explosion (BLEVE)
▶ Pressure burst
▶ Toxic gas release
▶ Fires, including pool fires, jet fires, flash fires and fireballs

The control room should be able to withstand overpressure in order to ensure that risks to personnel are acceptable. The design should incorporate safety measures, e.g. using laminated or polycarbonate glass in windows to protect against serious injury, and should consider using internal fixtures and fittings which are able to withstand the effect of a building shaking.

Should there be a risk of a release of toxic gases, the control room should be a place of safety where the ingress of toxic gases is controlled so that it cannot rise to a level which will affect the health and safety of the operators and/or make them unable to maintain control of the plant.

Careful consideration of the building's ventilation system is required to ensure that air intakes are situated away from areas that may be affected or to ensure that there is no air intake during an incident, preferably by closure of an automatic valve linked to a gas analyser.

The ventilation system should be designed to ensure that in the event of an incident occurring, the air intakes can be shut down. Ideally, this should be done automatically, i.e. having Heating, Ventilation and Air Conditioning (HVAC) duct inlet dampers installed. The design should also require that air intakes are placed away from areas which may be affected by incidents.

Control rooms should be constructed from materials which are fire resistant and able to withstand the effects of thermal radiation without the risk of collapse.

6.1.4 Temporary Refuge (TR) integrity

Temporary Refuge (TR) integrity can be described as the ability to protect the occupants following a hazardous event for a specific time period such that they will remain safe until they decide there is either a need to evacuate the installation, or to recover the situation.

This protection is applied by means of siting the temporary refuge in a position away from likely sources of fire, explosion or toxic gas release and being built to a standard that will withstand the effects of fire, explosion or toxic gas release. These will include building the TR of fire resistant material, building it to a blast resistant design and standard, and having temporary refuge Heating, Ventilation and Air Conditioning (HVAC) duct inlet dampers installed. These duct inlets are an obvious threat to the integrity of the temporary refuge and, as such, are required to be activated on any alarm detection of smoke, gas, vapour or heat.

A temporary refuge HVAC fire damper has two purposes:

1 To provide fire protection for the temporary refuge structure and heating, ventilation and air conditioning duct inlet and outlet ducting to the same standard as the fire rating of the wall it is penetrating.

2 To provide some sealing function to prevent excessive temporary refuge air outflow and the ingress of toxic smoke and gas for the specified temporary refuge endurance time – this is usually about 1 hour.

The design of a temporary refuge should take into account potential hazards and allow personnel access to a safe evacuation route. This means that, should a facility be evacuated, the route to safety should take personnel away from areas where plant and hazardous materials are stored as well as away from any potential exposure to toxic gases or fumes.

6.2 Management of change controls

6.2.1 Management of change

Many accidents and incidents, some of them catastrophic, can be attributed to changes in processes and equipment. Management should have systems in place to ensure that any proposed changes are evaluated before they are implemented. The management of change procedure should:

▶ Include expert personnel to review the proposed changes to ensure that they will not result in any operations exceeding established operating limits.

▶ Ensure that any proposed changes are subject to a safety review using hazard analysis techniques (e.g. hazard and operability studies) to assess the risks. This should be conducted by a team with relevant expertise which covers all areas of the process.

▶ Have in place arrangements for the control of relevant documents (e.g. safety case/report, drawings, Piping and Instrument Diagrams (P&IDs), etc.) and ensure that they are kept up to date. Any operating procedures should always be updated to reflect any changes in the plant or process.

▶ Ensure that any changes in the operating envelope (e.g. temperatures, pressures, flow rates, etc.) are communicated to the operators and documented. The operators should also have appropriate training to ensure competent and safe use of the equipment.

Temporary changes do need to be included in the management of change procedure, although it may not be necessary to include all the requirements of a permanent change. For example, master Piping and Instrument Diagrams (P&IDs) may not need to be changed although changes may need to be documented on temporary drawings.

REVISION QUESTIONS FOR ELEMENT 2 CONTINUED

2   When considering plant layout and the possibility of explosion, **explain** what mitigating factors can be introduced at the design stage.

3   Changes to processes and process operations can potentially introduce hazards, some inadvertently. **Explain** what measures can be introduced to ensure changes in procedures are managed as safely as possible.

You can compare your answers with our guidance at the end of the book under the section entitled 'Revision and assessment guide'.

CHAPTER 7

Element 2

Sub-element 2.3:
Role and purpose of a permit-to-work system

This chapter covers the following learning outcome

Explain the role and purpose of a permit-to-work system.

7.1 Role and purpose of a permit-to-work system

A Permit-To-Work (PTW) is a detailed document which describes specific work at a specific site at a particular time which is to be carried out by authorized personnel. It also sets out any precautions and control measures which are necessary to complete the work safely.

A PTW system is an integral part of a safe system of work and can help to properly manage a wide range of work-related activities. This is particularly true of maintenance work, where activities involve being in close proximity to normally restricted operations or energy sources.

Permit-to-work systems form an essential part of the controls that are forthcoming from a work task risk assessment process. When a work task is identified as being required, an assessment should be carried out to identify the nature of the task and its associated hazards.

From that assessment it should be possible to identify the risks associated with the task and determine the controls and precautions required that will mitigate the risks to an acceptable level. These controls may well include the need for a PTW system to be implemented.

A permit-to-work is not simply a means of giving permission to carry out a dangerous task; it is a means of defining how a task can be carried out safely and under what circumstances. Being a permit-to-work system, it also helps communicate the seriousness and importance of applying a prescribed method of working both to the person undertaking the task and to managers who control the permit-to-work system.

The issue of a permit-to-work does not, by itself, make a job safe – that can only be achieved by those preparing for the work, those supervising the work and those carrying out the work.

A PTW system on its own may not be enough to eliminate all risks associated with the task. Other precautions may be deemed necessary, such as process or electrical isolation or access barriers. These will need to be identified as part of the task risk assessments undertaken at the outset.

The PTW system allows for competent people to give consideration to the foreseeable risks of a particular task, so that such risks can be reduced to a level that can be regarded as being As Low As Reasonably Practicable (ALARP). Those carrying out the task also need to be competent and understand what they need to do in order to carry out their work safely.

7.2 The key features of a permit-to-work system

The essential features of permit-to-work systems are:

- ▶ The clear identification of who is responsible for specifying any necessary precautions
- ▶ The identification of personnel who may authorize particular jobs (including any limitations to their authority)
- ▶ The clear identification of any work classified as hazardous
- ▶ Clear identification of:
  - ▷ tasks
  - ▷ risk assessment
  - ▷ duration of permitted tasks
  - ▷ additional or simultaneous activity
  - ▷ control measures
- ▶ Training and instruction regarding the issue, use and closure of permits-to-work
- ▶ Monitoring and auditing to ensure that the system is working as planned

A permit-to-work document also provides written evidence, by means of a signature, that at various stages during the time the PTW was live, each stage was accepted by the person holding the PTW. This occurs when the PTW is issued from both the permit authorizer and the person accepting the permit; at handover points when work is transferred from one shift to another; at handback point when the work has been completed and ready for testing and re-commissioning; and at cancellation of the permit to certify the plant was satisfactorily re-commissioned and is safe to operate.

Let's now look at case study 7.1 which shows what can happen in a situation which clearly needed a permit-

Case Study 7.1

An oil storage tank farm had a redundant storage tank which needed to be demolished. It was sited near an operational pump house so it was agreed that the best way to conduct the work was to cold cut parts of the tank and remove them to a safe place for hot cutting into smaller pieces.

Contractors were engaged to undertake the work. Discussions between the client and the contractor took place to establish the agreed method of working. However, the reasoning behind the decision to cold cut the tank was not conveyed, and when it became apparent that cold cutting was becoming an arduous task, the contractor decided to use hot cutting methods. This ignited some flammable vapours which emanated from the nearby pump house and quickly developed into a serious fire.

During the subsequent attempt to quell the fire by the emergency services, five firemen were taken to hospital suffering from the effects of the fumes.

The subsequent enquiry concluded that a permit-to-work system should have been implemented and the work monitored by the client.

to-work system applying to ensure agreed safe work procedures were carried out.

A permit-to-work can also be described as a 'permit' or 'work permit', but all refer to a form or certificate which is used within a controlled system of work. It can be in a written or electronic form, whichever meets the needs of the company. It also provides evidence that consideration has been given to the risks of any job on site or any simultaneous and conflicting activities which may be affected by it.

The objectives and functions of a permit-to-work system are as follows:

▶ It ensures that proper authorization of designated work has been granted.
▶ It ensures those people who are conducting the work know the exact nature of the task including hazards, restrictions, time limitations, etc.
▶ It specifies the controls and precautions necessary to undertake the work safely, e.g. isolating machinery.
▶ It ensures those in charge of the location are aware the work is being carried out.
▶ It provides both a system of continuous control and a record that appropriate precautions have been considered and applied by competent persons.
▶ It affords the ability to display, to those who need to know, exactly what work is ongoing.
▶ It provides a procedural means of suspending work when this is necessary.

▶ It provides an ability to control work which might interact or conflict with ongoing operations or other permit-to-work activities.
▶ It provides a procedural means of handing over the work when that work covers more than one shift.
▶ It provides a procedural means of handing back the area or plant which has been involved in the work.

Let's have a look at case studies 7.2 and 7.3.

7.2.1 Appropriate use of Permit-To-Work (PTW) systems

Permits-to-work should be applied whenever work is to be carried out which may affect the health and safety of personnel, plant or the environment. This does not include work which is routine (e.g. routine maintenance which is carried out in non-hazardous areas) because this would tend to undermine the effectiveness of the permit system.

They should be considered, however, for:

▶ Special operations (i.e. work that is not routine)
▶ Work which is done other than normal production work (e.g. inspection, testing, dismantling, modification or adaptation of processes as well as repair work and non-routine maintenance)
▶ Work done by two or more individuals or teams where activities need to be co-ordinated to ensure the work is completed safely

7

Case Study 7.2

In 1992, maintenance work was being carried out on a still at Hickson and Welch Ltd in Castleford, West Yorkshire, in the UK. The work involved softening the residue with steam and then raking out the residue from the bottom of the still with a metal rake. During the operation, the action of the metal rake caused a spark which ignited the flammable atmosphere inside the still, which in turn caused a jet flame to erupt from the access opening and engulf the control room situated next to the still. As a result five people were killed.

The resulting investigation found a number of failings including:
• Failure to analyse the chemical content of the sludge prior to starting the job.
• Failure to analyse the atmosphere in the vessel prior to starting the job.
• Failure to control the temperature of the steam used to soften the sludge. The consequence of this was

that temperatures in excess of 90°C were applied, which affected the flammable range of the volatile atmosphere.
• The use of a metal rake in a flammable atmosphere.
• Failure to properly isolate the vessel prior to starting the job.

The investigation also found that, although a permit-to-work system was in place on the site and two permits were issued relating to this particular job, they only covered limited procedures. These were the removal of the lid from the still in order to the access opening and the blanking of the still inlet base. No permit-to-work was issued for the work conducted inside the still.

Had a permit-to-work been issued for the raking out of the residue then the system may have offered the opportunity to identify the hazards. Once identified, controls could have been applied to counteract the hazards.

Case Study 7.3

Work being carried out on the Piper Alpha rig at the time of the disaster was covered by a permit-to-work system. However, there was no cross-referencing of permits when work was conducted under one permit

which affected the work conducted under another permit. Reliance was placed on the memory of the designated authority.

▶ Work which will take longer to complete than one shift and when the work and responsibility for it needs to be formally handed over

7.2.2 Display

Copies of a PTW need to be clearly displayed:

▶ At the place where the work is to be carried out or in a recognized location close by. If the work is to be carried out in several locations, the permit should be kept by the performing authority.
▶ In the main control room or permit co-ordination room as well as in local control rooms.

As well as being displayed as described above, the issuing authority should keep a copy (unless that person is not located at the work site or in the control room, in which case the area authority should hold a copy).

Let's look at case study 7.4 relating to display of permit-to-work copies.

7.2.3 Suspension

Sometimes work has to be suspended, for example:

▶ If there is a change to the planned type or extent of work
▶ If there is a risk that the work to be carried out will present a hazard. For example, if the PTW is for hot work and at the same time there is a need to carry out sampling of process fluid or gas which itself poses the risk of the release of a dangerous substance.
▶ Whilst awaiting the delivery of spares.
▶ Where the work is in conflict with other work being carried out.

Although the permit-to-work is suspended, it will still remain live until it is actually cancelled. It is important, therefore, to remember that any isolations may still be active even though the work has been suspended.

Let's look at case study 7.5 relating to suspension of permits-to-work.

The permits which have been suspended should still be kept on the permit recording system. They should specify the condition of the plant where the work has been suspended and any consequences for other activities which may be affected by that condition.

If the suspension of work is likely to be extensive, it may be better to cancel the permit and re-issue a new permit when the work restarts.

7.2.4 Permit interaction

When a permit-to-work is being considered for a job, it is important that the person issuing the permit is aware of any other activity (either planned or already under way) which may interact with the work to be done under the permit.

There are many types of permit-to-work, and these may be issued by different issuing authorities even though they all apply to the same site or facility. Because of this, close liaison is required by the various authorities to ensure that work done under one permit does not interact with another, creating hazards for either of them. By cross-referencing permits, detailed information about activities which may interact can be identified (e.g. isolations on one job may interact with work on another).

Case Study 7.4

Again, referring to the Piper Alpha disaster, the subsequent inquiry found that the performing authority's copy of the permit was not always displayed at the job site, even though this was contrary to written procedures. More often than not it was kept in the performing authority's pocket. The subsequent report made a specific recommendation on this point:

'Copies of all issued permits should be displayed at a convenient location and in a systematic arrangement such that process operating staff can readily see and check which equipment is under maintenance and not available for operation.'

Case Study 7.5

Again, the Piper Alpha disaster presents us with an appropriate example. In this case, the control room, where the suspended permits should have been kept, was regarded as not having enough room to accommodate all of them. Consequently, they were kept in the safety office.

It was possible for a lead production operator to be aware of a suspended permit-to-work if it came to him for suspension in the 45 minutes before he officially came on shift. However, if the suspension had happened days earlier, or even on the same day before he came on shift, they could be completely unknown to him.

What made things even more difficult was that suspended permits were filed according to the trade involved rather than the location of work activity. The result was that it was not clear which equipment had been isolated in any one location.

The enquiry also found that there were often cases of large numbers of suspended permits at any one time. This made it difficult to establish which equipment was undergoing maintenance.

In the case of work done offshore, the control of conflicting activities by cross-referencing and prioritizing work tasks, to ensure that interaction between them does not create hazards, is usually the responsibility of a management team led by the Offshore Installation Manager (OIM). This team will prioritize work tasks during a permit-to-work meeting which will be held before any work is authorized.

7.2.5 Handover

A handover procedure is required in the event of any work being done under a permit-to-work which carries over into another shift. This allows the outgoing shift to communicate all relevant information about the work and the conditions of the permit to the new shift to ensure continuity and safety. The information will include:

▶ Any outstanding work under permit control
▶ The status of that work (this should be left in a condition which can be reliably explained to and understood by the incoming shift)
▶ The status of any other work which may affect the permit-controlled job (e.g. isolations)
▶ The status of the plant or installation

The information can be recorded in a permit log, permit file or on display boards. It is essential, however, that there is good communication between both the outgoing and incoming issuing and performing authorities. The signature of acceptance to allow the continuation of the permit by the incoming issuing authority is recommended.

Let's look at case study 7.6 involving handover as an issue.

7.2.6 Handback

The handback procedure should be a process of reinstating a plant or installation (or the part that has been worked on under a permit). This will ensure that the work has been completed and is in a safe condition. The issuing authority (the person who issued the permit) is responsible for signing off the permit and in doing so verifies that the plant or installation is in a safe condition, e.g. any isolations have been removed.

The person who is in control of operational activities will then make reference on the permit that the plant or equipment (or the part which has been subject to work under a permit) has been reinstated to the control of the production staff.

7.2.7 Permit authorization and supervision

A permit-to-work (PTW) must be co-ordinated or controlled by the issuing authority (or other responsible authority). This includes the monitoring and supervision of the work to ensure that it is being done according to the specific procedures detailed on the permit. Site visits by the issuing authority will depend on the work, the complexity and duration of that work and the hazards involved. The site visits should be carried out at least twice – at the start and completion of the work but more often if required. This will ensure that the conditions of the permit are being complied with.

7.2.8 Permit-to-work template

In Figure 7.1 we have reproduced a template of a permit-to-work document from the Health and Safety Executive (HSE) publication HSG250, which states what information is required, a description of that information where necessary and the authorization required where necessary.

7.3 Types of permit

Permit-to-work (PTW) covers many different types of operations and tasks, and the following are examples of types of job where permits should be considered:

▶ Work where heat is used or is generated, for example welding, grinding, etc.
▶ Work which involves breaking containment of a flammable or dangerous substance
▶ Work which involves breaking containment of a pressure system
▶ Work on electrical equipment
▶ Work within tanks and other confined spaces
▶ Working at height
▶ Work involving hazardous substances
▶ Well intervention
▶ Diving operations
▶ Work involving pressure testing

Let's have a look at some specific types of permit.

Types of permit – hot work permit

This is issued for work which involves the application of heat or sources of ignition to vessels or equipment which may contain or have contained flammable vapour. Also for areas in which there may be a flammable atmosphere.

Case Study 7.6

The Piper Alpha inquiry found that during the handover on the night of the disaster, operators failed to communicate the fact that a service valve had been removed for servicing and had not been replaced. This missing service valve was the source of the leak of hydrocarbon which subsequently ignited and caused the disaster.

Permit-to-Work	
1) Permit Title	2) Permit Number Reference to other relevant permits or isolation certificates.
3) Job location	
4) Plant identification	
5) Description of work to be done and its limitations	
6) Hazard identification Including residual hazards and hazards associated with the work.	
7) Precautions necessary People who carried out precautions, e.g. isolating authority, should sign that precautions have been taken.	
8) Protective equipment	
9) Authorization Signature (issuing authority) confirming that isolations have been made and precautions taken, except where these can only be taken during the work. Date and time duration of permit. In the case of high hazard work, a further signature from the permit authorizer will be needed.	
10) Acceptance Signature confirming understanding of work to be done, hazards involved and precautions required. Also confirming permit information has been explained to all permit users.	
11) Extension shift handover procedures Signatures confirming checks made that plant remains safe to be worked upon, and new performing authorities and permit users made fully aware of hazards/precautions. New expiry time given.	
12) Handback Signed by performing authority certifying work completed. Signed by issuing authority certifying work completed and plant ready for testing and re-commissioning.	
13) Cancellation Certifying work tested and plant satisfactory re-commissioned.	

Figure 7.1 Permit-to-work template
Source: Wise Global Training interpretation based on HSE publication HSG250.

Hot work permits are typically coloured red or are red-edged.

Hot Work Permit

Figure 7.2 Hot work permit
Source: Wise Global Training.

Types of permit – cold work permit

This is issued for work involving hazardous activities which are not covered by a hot work permit.

Cold Work Permit

Figure 7.3 Cold work permit
Source: Wise Global Training.

Types of permit – equipment disjointing certificate/breaking containment permit

This permit is required for any work which involves the disconnection of pipework or equipment which may contain (or has contained) any high pressure or hazardous fluids or other substances. It will usually be used for the insertion of spades into pipework (and for the subsequent removal of those spades). These certificates usually have a black border.

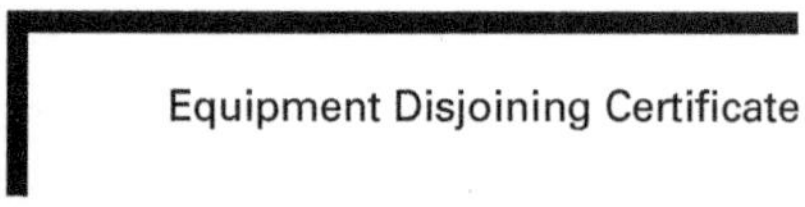

Equipment Disjoining Certificate

Figure 7.4 Equipment disjointing permit
Source: Wise Global Training.

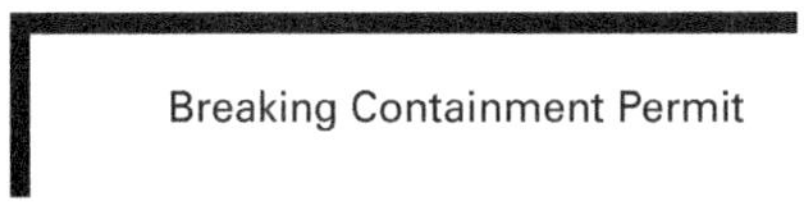

Breaking Containment Permit

Figure 7.5 Breaking containment permit
Source: Wise Global Training.

Types of permit – electrical work permit

As it suggests, this permit is used when working on a piece of equipment or a circuit that is safe. A permit should never be issued for work on live equipment.

Types of permit – confined spaces entry certificate

These certificates are used when entry to a confined space is essential for work to be done. They should specify all of the precautions necessary to ensure that exposure to hazardous fumes or an oxygen-depleted atmosphere is eliminated before entry to the confined space is permitted.

Although the certificate should confirm that the enclosed space is free from asphyxiating gases or hazardous fumes, it should also specify any precautions necessary to protect the worker(s) from exposure to the risk of harm from other sources, e.g.:

▶ The ingress of airborne contaminants from other sources
▶ Hazardous fumes being released from residues within the confined space
▶ Oxygen depletion caused by oxidation

These precautions can include:

▶ Use of forced ventilation
▶ Provision of personal protective equipment including breathing apparatus

Types of permit – working at height permit

A working at height permit would be used to specify precautions required to fulfil the following requirements:

▶ Protection for the person who will be working at height (fall arrest equipment)
▶ Precautions for the safe rescue of a person should he/she fall
▶ Supervision of the worker to ensure safe working procedures are being followed
▶ Protection from falling objects
▶ Safe access to and egress from the area

Types of permit – machinery certificate

A machinery certificate would be issued for work to be done on large, complex pieces of equipment, and this would specify procedures to ensure the correct isolation of that equipment before the work commences.

Types of permit – isolation certificate

As with an electrical or machinery certificate, an isolation certificate is used to ensure that any equipment is mechanically and electrically isolated before work commences. A similar certificate may be used to confirm chemical isolation of plant and machinery. All of these certificates, if used, should be cross-referenced.

Types of permit – radiation certificate (non-destructive testing of welds)

Radiation certificates are issued to specify control measures to minimize the risk of exposure to sources of radiation. These control measures include:

▶ Site inspection
▶ Controls on sources of exposure
▶ Radiation monitoring
▶ Access or containment barriers

The certificate should include specific requirements for:

▶ Access/egress to/from the site
▶ The minimum number of people needed on site during the radiography procedure
▶ Special emergency procedures for radioactive source transportation

Types of permit – diving certificate

A diving certificate is used to ensure the safety of the dive team. It ensures that no other work activities are taking place at the same time which would put the team at additional risk. This could include any over-side work or live firewater intake pumps.

7.4 Interfaces with adjacent plant

As we have already mentioned, any permits for work being considered must be cross-referenced to any other work either being planned or in progress. This is to ensure there is no conflict or interaction which may result in the creation of additional hazards for any of the work.

For example: a maintenance crew is working on machinery on the main deck, a task which is scheduled to take one day. However, unforeseen problems arise and the work runs into day two.

On day two, as pre-arranged, a team of scaffold erectors join the rig to start erecting scaffold on a deck above the machine having maintenance carried out upon it. The scaffolding work, which if it went ahead would be conducted over the maintenance team, would clearly present a risk of injury to the members of that team.

A permit-to-work (PTW) system which cross-references other work would be expected to identify this conflict of work before it came about and the scaffolding work would be delayed until deemed safe to proceed (i.e. when the maintenance crew had finished their work).

7.5 Interfaces with contractors

Contractors, as we have covered earlier in Chapter 5, are those people who have been engaged to undertake work for a client but who are not employees of the client. Consequently, they may not be as familiar with hazards, working practices and procedures, and permit-to-work (PTW) requirements, as are company employees. This unfamiliarity puts them at far more risk of harm.

There are a number of possible reasons why accidents which involve contractors are more prevalent, for example:

▶ Unfamiliarity with the layout of the site
▶ Unfamiliarity with the permit-to-work (PTW) systems in place at the client's site
▶ Lack of adequate training for the contractor about the implementation of the permit-to-work procedures for a particular site
▶ Lack of communication or knowledge within the workforce of the presence of the contractor
▶ Increased hazards associated with the specific type of work which the contractor provides

Consequently, a comprehensive induction procedure should be undertaken with regard to permit-to-work (PTW) systems with all contractors prior to any work being undertaken. This should include:

▶ Ensuring the contractor understands fully the principles of the PTW systems within the industry
▶ Ensuring the contractor and his/her employees fully understand the PTW systems and the arrangements within them which apply to the site where they are to carry out the work
▶ Ensuring that all personnel from the performing authority and other users are fully trained and aware of any specific arrangements in force to make the job safe at the location where they are to carry out the work

The following issues should be considered in relation to the issuing of permits to contractors as far as interfaces are concerned:

▶ The client has to be certain that his/her employees are not exposed to any risk due to the actions of the contractor or his/her employees.
▶ The client must ensure that his/her employees do not expose the contractor or his/her employees to any risk due to their actions.
▶ The client must give the contractor full information regarding any hazards and risks associated with any of the parts of the work area where the contractor or his/her workforce will be undertaking their work. An example of this would be where contractors would be working close to hot steam pipes.

▶ A client must ensure that the contractor and his/her workers work in accordance with agreed safe systems of work, and to do this he/she must monitor the work during the period of the contract.

7.6 Lock out, tag out and isolation

7.6.1 Introduction

The term 'lock out' or 'tag out' (LOTO) refers to a procedure which is aimed at safeguarding someone who is working on or near plant or machinery, from that plant or machinery unexpectedly starting up or releasing energy of some kind.

This procedure involves an authorized person to disconnect the plant or machinery from its energy source. This person then locks or tags the isolator in order to prevent anyone from re-energizing the plant or machine.

A lock out device is a physical restraint which maintains the isolation device in a safe or 'off' position. It can only be removed with a key, or some other unlocking mechanism, which can only be used with the authority of the person in charge of the permit-to-work (PTW) system.

For safety reasons, each lock must only have one key. No master keys are allowed on a lock out system. We have shown examples of lock out and tag out devices below. Figure 7.6 shows an example of a lock out device fitted to an isolator which prevents the isolator from being switched on.

Figure 7.6 Lock out device
Source: iStock.

A tag out device is more of a warning system that an authorized worker applies to the isolator in order to indicate work is being carried out on the plant or machine. As such, the isolator should not be switched on whilst the work is underway.

Tag out devices are easy to remove. Consequently, they offer less protection than lock out devices. Figure 7.7 shows an example of a tag out device.

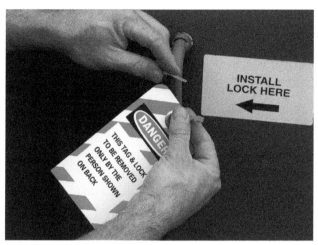

Figure 7.7 Tag out device
Source: iStock.

7.6.2 Basic procedure for implementing lock out and tag out actions

1 The person authorized to conduct the lock out or tag out action will identify the source(s) of energy to be controlled and the method of control to be used. A safe working procedure will then be established to implement the isolation followed by the lock out, tag out action.
2 All personnel who may be affected by the isolation caused by the lock out, tag out action should be informed. This information will include:

   ▶ what plant/machinery is going to be isolated
   ▶ why it is going to be isolated
   ▶ how long it will be isolated
   ▶ who is the authorizing person
   ▶ who is the contact person .

3 The process or system should be shut down as normal and confirmation made that all controls are in the off position and that all moving parts have stopped.
4 The process or system should be isolated from its source of energy, following the safe working procedure set out by the authorized person at the identification of energy stage. Depending on the source of energy, the procedure will vary as follows:

   ▶ Electrical energy – the isolator to be switched to off position and the breaker connections to be visually verified to be in the off position. The isolator switch to be locked in the off position with a lock out or tag out device.
   ▶ Hydraulic or pneumatic energy – the isolator valve should be set in the closed position and locked in the closed position with a lock out or

tag out device. Pressure should be bled off by opening the pressure relief valve.

   ▶ Mechanical energy – energy should either be carefully and slowly released from compressed spring, or the components should be blocked in the system or process which has the potential to be moved or activated by the spring's energy. If the components are blocked from moving, the block should be locked in position with a lock out or tag out device.
   ▶ Gravitational energy – an isolation block or pin should be positioned to stop the system or process from moving. The block or pin should be locked in position with a lock out or tag out device.
   ▶ Chemical energy – the isolator valve on the supply lines should be set in the closed position. The isolator valve should then be locked in the closed position with a lock out or tag out device. Where possible, or appropriate, chemicals should be bled off by opening the bleed valve.

5 Verification that the isolation is effective needs to be made. This is normally done by making an attempt to try to restart the system or process. If it does not start, the isolation has been effective. Once isolation is verified the start-up controls must be returned to their off or neutral position.
6 Once the work is completed, the lock out, tag out device will need to be removed. This operation should include:

   ▶ Ensuring all tools and equipment have been removed
   ▶ Ensuring all personnel are not near any hazardous areas
   ▶ All controls are in the off or neutral position
   ▶ Removal of the lock out, tag out devices and re-energizing the plant or process
   ▶ Informing relevant personnel that the plant or process is back in service
   ▶ The person who has been undertaking the maintenance work should remain whilst the plant or process is restarted

REVISION QUESTIONS FOR ELEMENT 2 CONTINUED

4 **Explain** the role and purpose of a permit-to-work system.
5 **Outline FOUR** objectives of a permit-to-work system.
6 **Describe** the basic procedure for implementing a lock out/tag out action.

You can compare your answers with our guidance at the end of the book under the section entitled 'Revision and assessment guide'.

Sub-element 2.4:
Key principles of safe shift handover

8.1 Placing greater reliance on written communication between handover of 12-hour shifts

8.1.1 Shift handover – a definition

Where a production process or operation continues indefinitely, there inevitably comes a time when those personnel undertaking or overseeing the process or operation have to leave their duties and let others take their place. This is when 'shift handover' takes place.

However, when that process or operation involves complex issues and/or hazardous substances, it's crucial that all information appertaining to the period of the outgoing shift is conveyed to those who are taking over so that continuity can be maintained and safety upheld. Shift handover typically includes:

1 A period of time where the outgoing team prepares the information it will be conveying to the incoming team

2 A period of time where both the outgoing and incoming teams communicate with each other and exchange all relevant information

3 A period of time where the incoming team cross-checks the information passed onto it as it takes on the responsibility for ongoing operations

The main aim of safe shift handover is to transfer accurate, reliable and relevant information from the outgoing shift team to the incoming shift team in order that there is continuity of safe and effective operations.

8.1.2 Shift handover – background

Oil and gas exploration and production processes are continuous 24-hour operations. The personnel who work on these operations and processes tend to work alternate 12-hour shifts over a 2–4 week period.

These operations tend to be highly complex and place demands on the operator's skills as far as information processing and decision-making is concerned. Also, processes may evolve over many hours, or even days, and the ability of the operator to clearly understand what is going on, or what is expected, is critical. This mental appreciation of the situation is one of the more challenging issues when it comes to shift handover.

Successful control of an operation requires three elements:

1 A clear understanding of the expected outcomes of the production process or operation
2 A clear understanding of the current state of the production process or operation
3 A clear understanding of the dynamics of the production process or operation

As we've just mentioned, many processes evolve over a long period of time. Tasks undertaken to adjust a process may take hours or days to take effect. Consequently, good communication at shift handover on expected long term or delayed reactions is essential.

8.1.3 12-hour shift handover

In the 1970s, shift patterns in the oil and gas industry gradually changed from 8-hour to 12-hour shifts. The view at the time was that inter-shift communications improved because the number of handovers was reduced, which in essence meant the number of occasions when critical information might not have been passed over was effectively reduced. This awareness highlighted the fact that quality of handover was a significant issue and needed to be addressed.

Shortly after, a research based project was conducted at an oil refinery in the UK where personnel involved in process shift work were asked to set out exactly the kind of information they would need to have passed to them in order to undertake their work safely and effectively if they were the incoming shift workers.

Once this information was collated, it was used as a design template for a structured log book. The result was that this log book was used as a constant record of events and activities which occurred during a shift. At handover, it was used as a prompt to instigate conversations as to the information needed to be discussed between outgoing and incoming staff. It was generally accepted that talking through the log book afforded the handover process a more structured approach, giving greater continuity between shifts, more information being passed on, and key information being recorded in writing and discussed verbally.

Furthermore, the project involved conducting interviews with a wide range of experienced staff to establish effective handover procedures. The analysis of these interviews, together with the structured log book, led to a written procedure being established on how to conduct an effective shift handover.

One thing which did become apparent from the project was that handover should not be restricted to a limited time. Indeed, handover should take as long as necessary.

8.2 Two-way communication at handover with both participants taking joint responsibility

During shift handover, the communication of information from one person to another is not just a one-way process. The person who is providing the information needs to convey it in a way such that the person who is receiving it is in no doubt about the message being conveyed. It is a mutual interaction between the two people – person A (the sender of the information) and person B (the person receiving the information). Person A needs to receive feedback from person B confirming that the information has been received and correctly interpreted in order for the two-way communication to be successful.

Studies of social interaction within small groups have provided some evidence that highlights the need for such feedback – it is evident that increased feedback

Case Study 8.1

Following the Piper Alpha disaster, the inquiry concluded that one of the main contributory factors to the disaster was a failure to convey critical information at shift handover. In particular, a pressure safety valve had been removed and had been replaced by a blind flange. The incoming shift, being unaware of this fact, took actions which to their mind were normal procedures but which actually triggered the disaster.

The report which followed the inquiry concluded that there was no established procedure as to how shift handover should be conducted or what should be included.

is associated with increased accuracy and confidence in communication. The increase in feedback, however, does mean that more time needs to be allowed for in the communication process.

A recent theory on reliable communications emphasizes the role of feedback in accurate communication and also suggests a possible reason why, under certain conditions, shift handover may be problematic. The theory is based on an analysis of accidents and the concept of mental models (a mental model being the internal mental representation which is held by an individual). In the context of the oil and gas industry this could be the mental model of the installation that individual is responsible for operating.

According to this theory, two factors should be considered when assessing the reliability of communications:

1 The mental models held by those persons communicating with each other. If those models are compatible, e.g. under normal plant conditions and between experienced operators or when both parties have been on duty for several consecutive shifts, then the handover will generally be unproblematic. This is because they both have a shared understanding of the information being conveyed.
On the other hand, when the mental models are not compatible, e.g. under abnormal plant conditions and between experienced and inexperienced operators, or if there has been a significant break between shifts, then the shift handover could be problematic. This means that good communication is necessary to enable both mental models to become compatible.

2 Feedback is the second factor essential for effective communication. Under normal circumstances when the mental models are compatible and there is a shared understanding of the information, feedback is not as essential because there is less risk that information being conveyed will be misinterpreted.

Under abnormal conditions, feedback is more important to ensure effective communication as it allows:

▷ the receiver of the information to confirm that he/she has received, interpreted and understood what has been communicated
▷ the sender to confirm that the information has been successfully communicated as well as giving him/her the opportunity to clarify any issues if necessary

On conveyance of information, the receiver registers his/her understanding of the information.

On conveyance of understanding by the receiver, the sender knows that the receiver has registered and understands the sent information.

8.3.1 Why good shift handover is important

Good shift handover allows the accurate and effective communication of vital information to incoming personnel in order to enable the safe operation of process plant and equipment.

The aim of shift handover is to communicate effectively all relevant information to allow the continuation of safe and efficient systems and process operations. As we mentioned before, there are three elements which contribute to an effective shift handover:

▶ A period of time where the outgoing team prepares the information it will be conveying to the incoming team.
▶ A period of time where both the outgoing and incoming team communicate with each other and exchange all relevant information.
▶ A period of time where the incoming team cross-checks the information passed onto it as it takes on the responsibility for ongoing operations.

A definition of an effective shift handover is:

The effective transfer of information between the outgoing and incoming parties with no miscommunication or misunderstanding.

8.3.2 Key principles in effective shift handover

Although the actual shift handover is between those ending their shift and those about to start their shift, management should also ensure that the facilities, the arrangements and the infrastructure allow for good shift handover practice. In effect, management should:

▶ Provide all staff who are involved in the handover process with training and the development of good communication skills
▶ Provide written procedures for effective shift handover
▶ Emphasize the importance of effective shift handover
▶ Ensure that both parties take joint responsibility for conducting an effective shift handover
▶ Emphasize the use of both written and verbal means of communication
▶ Ensure that any important information is written into a log or report form. The log or report form should be brought to the attention of personnel at handover.
▶ Include all personnel from both the incoming and outgoing team in the handover without exception
▶ Ensure that managers and supervisors are available at all times and implement an 'open door' policy

Shift handovers should involve the following principles:

- Be treated as high priority
- Not be rushed but be allowed as much time and resources as is necessary to ensure the accurate communication of information
- Be conducted using both verbal and written means of communication
- Be conducted face to face, with both parties taking joint responsibility for the effective communication of necessary information
- Be conducted in an environment which is conducive to good communication without distractions
- Involve all shift personnel

As far as the operation of the plant and equipment is concerned, issues covered should include:

- Work permits – the status of existing permits and the status of work in progress
- The updating of work permits
- Preparations for upcoming maintenance
- New personnel to the shift
- Any plant overrides – existing and planned
- Information about any abnormal events
- Any existing or planned shutdowns
- Any changes in plant parameters
- Any routine operations and existing parameters which may need to be carried out by personnel from the incoming shift
- Any breakdowns which may have occurred
- Any faults which have occurred with safety critical equipment
- Inhibits to the Fire and Gas (F & G) and Emergency Shutdown (ESD) systems
- Any completed work and equipment which has returned to service

REVISION QUESTIONS FOR ELEMENT 2 CONTINUED

7   **Outline** what the process of shift handover involves.
8   **Identify FOUR** of the principles involved in a good shift handover process.

You can compare your answers with our guidance at the end of the book under the section entitled 'Revision and assessment guide'.

CHAPTER

9

Element 2

Sub-element 2.5:
Plant operations and maintenance

This chapter covers the following learning outcome

Explain the importance of safe plant operation and maintenance of hydrocarbon containing equipment and processes.

9.1.1 Asset integrity – an introduction

In the oil and gas industry, offshore installations, refineries, storage terminals and pipelines are often referred to as 'assets'. Maintaining an asset's ability to function effectively and efficiently without creating undue hazards to any persons or the environment is essential and this is referred to as maintaining an 'asset's integrity'.

Managing an asset's integrity involves ensuring the people, the systems, the processes and the resources are all available and used when required in order to maintain integrity.

The management of an asset's integrity will cover its entire life from initial design through to commissioning and operating and, finally, to decommissioning. It will start with the designers, then the builders, manufacturers, operators, maintenance team, management, decommissioning planners and finally the decommissioning and demolition team.

Asset management also involves identifying all possible threats to the asset's integrity by looking at all conceivable scenarios. There are various methods that can be used to evaluate the threats to an asset, including undertaking a Hazard Identification (HAZID) study, which we covered in Chapter 3.

It's worth noting that, following the inquiry into the Buncefield explosion, the subsequent report recommended that the definition of the phrase 'conceivable scenarios' should be broadened to include less likely, and previously discounted, scenarios.

The day-to-day maintenance of managing an asset's integrity involves ensuring the following aspects are continually undertaken:

▷ Inspection
▷ Testing
▷ Maintenance
▷ Corrosion prevention
▷ Monitoring
▷ Competency
▷ Training

9.1.2 Safety Critical Elements (SCEs)

Because of the raised risk level of working with highly flammable, hydrocarbon based substances, it is essential that safety devices are incorporated within an asset so that if a hazardous situation or incident does occur, there are devices in place to deal with the consequences by quelling, controlling or mitigating the event or situation. These are known as Safety Critical Elements (SCE). They include:

▷ Blowout preventers
▷ Fire deluge protection systems
▷ Emergency shutdown valves
▷ Fire and gas detection systems

We will now take a brief look at each of these elements to understand the importance and significance of each of them and, as such, why their continued operability is essential.

Safety critical elements – blowout preventers

Figure 9.1 Blow out preventer
Source: iStock.

Blowout Preventers (B-O-Ps), as their name suggests, are used to control blowouts. When a blowout occurs, i.e. there is a sudden surge of pressure from within the drill hole, the B-O-P is automatically activated by the ensuing pressure, ensuring that the pressure is contained and the hole is sealed. As can be imagined, without such a device, the consequences can be catastrophic. You only have to look at what happened when the *Deep Water Horizon* incident happened in the Gulf of Mexico in 2010. This was the result of a malfunctioning blowout preventer.

Figure 9.2 Fire deluge system
Source: Frontline Fire International.

Consequently, B-O-Ps should be regularly maintained, inspected and tested.

Safety critical elements – fire deluge systems

Fire deluge systems are positioned where the risk of fire is a serious concern. Figure 9.2 shows a fire deluge system in action at a road tanker loading area. They can apply a continuous and high volume of water to an area which includes the hazard as well as possible escape routes for personnel.

They work by having a number of water outlets aiming at the desired target area. These outlets are kept in the open position but not supplied with water until the pumps that supply the water are activated by a fire sensor.

Again, regular maintenance, inspection and testing are essential requirements. This is because blockages to the pipes will render them ineffective. Blockages can be caused by simple things like corrosion and, if they use sea water as some deluge systems do, marine growth.

Safety critical elements – Emergency Shutdown Valve (ESDV)

An Emergency Shutdown Valve (ESDV) is a device which is designed to automatically shut down the flow of a fluid when a dangerous situation is detected. As its name suggests, when such a situation is detected, it automatically shuts down the process system and thus retains the integrity of the asset.

Emergency shutdown valves must operate within specific parameters. That is to say, they must close completely within a specified time. They must also operate within allowable specified leak rate limits when in the closed position. All these aspects need to be checked and tested regularly as well as being maintained to an agreed schedule.

Safety critical elements – fire and gas detection systems

As the name suggests, the fire and gas detection systems use two main types of detection devices, one type for fire detection and the other type for gas detection. The fire detectors are used to detect heat, flames and smoke, whilst the gas detectors are used to detect flammable and toxic gas as well as vapours.

Since they are the first line of defence in the case of a fire, gas or vapour escape, their ability to perform as expected is essential. All detectors, and the systems they are part of, should be regularly tested and maintained. The testing procedure should not only include testing to see if the instruments work, but should also include the ability of the devices to detect incidents wherever they may occur. This will determine if there are any 'dead' areas – areas where fire or gas escape may go undetected, at least for a while.

Safety critical elements – safety inspection and testing

All safety critical elements are the first and mainline defences in the case of a hazardous event occurring. If these elements fail to perform their role or function, the hazardous event may well escalate into a catastrophic

Figure 9.3 Emergency shutdown valve
Source: iStock.

incident. Consequently, it is essential that all safety critical elements are 100 per cent effective at all times, and the only way to be assured of this is to conduct a pre-defined programme of inspection, testing and maintenance.

For example, Emergency Shutdown Valves (ESDVs) must close within a specified time once activated and their leak rate must also be verified to be within acceptable limits.

Pipework and vessels which have been repaired or replaced should be pressure tested prior to being put back into service.

Computer programs and systems should also be tested on a regular basis.

Safety critical elements – corrosion prevention

Corrosion can deteriorate equipment to such a level that it can affect the integrity of an asset with potential catastrophic consequences. However, it is possible to predict corrosion rates in specific materials, and this information can assist in determining how often inspection and cleaning should be undertaken.

One of the biggest problems with corrosion is that it tends to occur in places which are difficult or even impossible to see with the naked eye, e.g. inside equipment or pipework. Fire deluge systems are typical in this respect. They should be tested and maintained on a regular basis.

If a fire deluge system draws its source of water from the sea then the maintenance programme needs to be

more frequent and robust. This is because salt water accelerates rates of corrosion as well as having the potential to introduce marine growth within the system, which may also be a cause of blockage.

When it comes to detecting and monitoring corrosion, there are a number of specialized computer based monitoring systems available, such as electrochemical noise and electrical resistance monitoring, which have been developed for use in areas prone to attack by corrosion.

There are also other inspection systems available to monitor potential corrosion, including Magnetic Particle Inspection (MPI), Non-Destructive Testing (NDT) and radiography, as well as visual inspection. These can also be used to monitor the effectiveness of corrosion prevention methods.

When it comes protecting material from corrosion, the use of cathodic protection systems, which include the use of sacrificial anodes, is a primary example, as is injecting chemical corrosion inhibitors into pipework and process systems.

Corrosion inhibitors are normally distributed from a centrally located inhibitor distribution skid. They are effective against corrosion as they:

▶ Increase the electrical resistance of metal surfaces
▶ Reduce the diffusion of ions to the metal surfaces
▶ Increase the cathodic polarization

Other methods of corrosion prevention include:

▶ The use of appropriate materials for the requirements of the job. This depends upon the level of anticipated corrosion, i.e. in areas where corrosion is minimal it may be acceptable to use carbon steel, which is not as expensive as materials which may be required in areas of potentially heavy corrosion. Other components may need to be made from stainless steel, which is more expensive, or from aluminium.

▷ Using protective coatings on the surfaces of metal equipment to prevent corrosion.

▷ The use of a veneer of expensive metal over a cheaper metal to provide protection (instead of the use of the full thickness of expensive metal).

Safety critical elements – training and competency

It is essential that personnel have the appropriate training in order to become competent in the skills necessary to maintain asset integrity. Equally as important is that personnel undertaking these maintenance tasks understand the importance of their work in retaining the integrity of the asset.

'Competence' can be defined as the ability to undertake responsibilities and perform activities to a recognized standard on a regular basis. It is a combination of skills, experience and knowledge.

Lack of competency has been a contributory factor in a number of disasters, including ESSO Longford and BP Texas City. It's poignant to note that the Offshore Installation Manager (OIM) on the Piper Alpha rig on the night of the disaster had not undergone any formal training in the management of major emergencies. Consequently, he was unable to manage the emergency with any level of competency as it unfolded.

In order to attain an acceptable level of competency, people first have to be trained to a level whereby they can efficiently undertake the tasks for which they have been trained.

Training can be described as a learning process which helps people to learn how to undertake a task or tasks. When that learning reaches a particular level of understanding and it is supplemented with regular on-the-job practical application of that knowledge, it gradually evolves into a skill. Further on-the-job application of knowledge adds a level of experience which can eventually be regarded as competence.

There are some groups of people who have particular training needs. For example:

▷ New or inexperienced personnel will need induction training (how to work safely, fire and evacuation arrangements, and first aid). They should also be adequately supervised until they become more experienced.

▷ Existing personnel who may be changing jobs or taking on extra responsibilities (they need to know what the implications are and how to deal with new health and safety issues).

▷ Personnel who may have been in the job for a period of time and need refresher training in order to maintain their skills and competence.

▷ Young people are more likely to be affected by accidents due to their inexperience in the world of work as well as in their particular jobs. They should be given priority in the provision of training and adequate supervision until they become competent.

As well as giving people new skills and knowledge, training can also:

▷ Give them self-confidence
▷ Improve their morale
▷ Give them the confidence to share their knowledge and skills with others
▷ Improve their attitudes towards health and safety
▷ Put a value on the work they do

Whatever the reason for training, whether it is for professional development or to embrace a specific skill, it also allows that person to conduct their work more effectively.

9.2 Risk based maintenance and inspection strategy

9.2.1 Risk based maintenance

Oil and gas facilities, whether they are offshore or onshore, involve working with highly flammable and/or toxic materials. The hazards associated with these materials are sometimes compounded by the extreme process conditions that have to be applied, such as high temperature and high pressure. Consequently, proper maintenance of process equipment is essential in order to ensure the safe and continuous operation of the facility. This means implementing an appropriate system of maintenance which will ensure hazardous situations are prevented from developing from defective process equipment.

To this end, Risk Based Management (RBM) techniques are used as a means of developing detailed maintenance schedules. These schedules aim to pre-empt faults occurring as well as deal with predictable normal wear and tear, and aim to provide a safe and fault-free operation of the process facility.

Risk based maintenance – overview

Risk based maintenance is based on increasing the reliability of equipment by assessing the probability of various failure scenarios and applying an appropriate maintenance schedule in order to pre-empt these failures.

In the first instance, the likelihood of equipment failure scenarios is formulated. Then, out of the many likely failure scenarios, the ones which are most credible (most likely to happen) are subjected to an in-depth

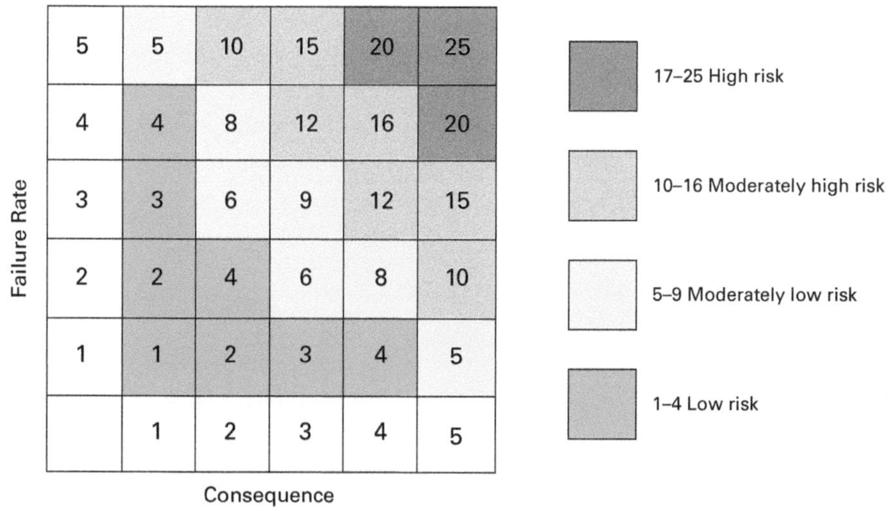

Figure 9.4 Risk matrix
Source: Wise Global Training.

analysis, including building a detailed consequence picture of each scenario.

From there, a fault tree analysis is drawn up to determine the probability of failure.

The risk of failure is calculated by taking the probability analysis results and combining them with the consequence analysis results.

Figure 9.4 is an example of a risk matrix. Once the risk has been calculated, it is compared with known acceptable criteria. The frequency of each maintenance task is then guided by the acceptable criteria in order to reduce the level of risk to an acceptable level.

Risk based maintenance – inspection

In order to ensure the integrity of plant and equipment during its service life, a regulated inspection strategy should be implemented to establish that:

▶ No damage has occurred to the plant or equipment
▶ Wear and tear are being kept under review and any components that are considered to be in need of replacement are notified to the maintenance crew

The inspection strategy is implemented for each piece of plant or equipment in accordance with the outcome of the fault tree analysis mentioned above, which takes into account the implications of structural failure. By implications we mean giving consideration to both the probability of structural failure and the consequences of such a failure.

By taking account of both the probability and the consequence of any failure, an inspection strategy demonstrates, in an auditable manner, that inspections have been conducted effectively and efficiently.

Risk based maintenance – maintenance

Plant and equipment have to be maintained so that they remain in a safe and reliable condition. Effective maintenance will involve knowledge of previous failures as well as maintenance records which may, when analysed, indicate trends of wear and tear.

Risk based maintenance follows the risk based inspection strategy we have just covered. It evolved as a means of achieving tolerable risk levels with plant and equipment that carry high risk.

9.2.2 Condition monitoring maintenance

Condition monitoring involves taking readings of equipment in ways that allows those readings to give an insight into the condition of the plant or equipment without having to shut it down or dismantle it. For example, equipment which has rotating components can be monitored using vibration analysis. This technique involves taking readings of the vibration patterns of critical components such as shafts and bearings. Analysis of these readings will show if there are any adverse aspects which require further work or more detailed inspections. This non-intrusive technique allows decisions on maintenance to be made more frequently without great disruption, which can result in a substantial reduction in the amount of maintenance needed on equipment. The tendency to over-maintain is also reduced.

9.3 Techniques, principles and importance of safe operation, standard operation procedures and maintenance

The hazards associated with the oil and gas industry have changed very little over the years. However, over that same period of time knowledge and understanding of these hazards by those within the industry has accumulated significantly and been greatly enhanced.

Furthermore, because everybody understands the high level of risk shared by all within the industry, this knowledge and understanding is actively shared as lessons are learned.

The requirements of the offshore industry in terms of safety are greater than their onshore compatriots. This is partly because of the harsh environments the offshore sector has to endure, plus the fact that operators on offshore facilities cannot simply walk away to a safer area or expect the emergency services to arrive within minutes.

9.3.1 Pre-start-up safety reviews

Apart from the reactive safety elements (emergency shutdown systems) built into any process system, there is also a need for proactive elements to be applied to operating procedures. An example of such an element is a pre-start-up safety review.

A pre-start-up safety review is conducted:

▷ Before the actual start-up of a new installation
▷ When new chemicals or other hazardous materials are introduced into a process
▷ When existing facilities have had significant modifications or a maintenance shutdown

The review is conducted to ensure that:

▷ All materials and equipment which have been used to construct/modify/maintain the process system conform to the design criteria.
▷ All of the following have been inspected, tested and certified:
  ▷ process systems and hardware (including computer control logic)
  ▷ alarms and instruments
  ▷ relief and safety devices and signal systems
  ▷ fire protection and prevention systems.
▷ Safety, fire prevention and emergency response procedures have been formulated, reviewed and tested and are appropriate and adequate.
▷ Start-up procedures are in place and all appropriate actions have been taken.
▷ A process hazard analysis has been conducted and any resulting recommendations have been implemented or resolved and any actions taken have been documented.
▷ All training (initial, induction and/or refresher) has been undertaken for all personnel and that it included training in:
  ▷ emergency response
  ▷ process hazards
  ▷ health hazards.
▷ All procedures have been completed and are in place, including:
  ▷ operating procedures (both normal and upset) and operating manuals
  ▷ equipment procedures

▷ maintenance procedures – finalization of maintenance procedures will include ensuring all pipes and drains opened during maintenance activities are closed off as required, all instrumentation has been replaced, calibrated and in good working order, and that all spades, blinds or spectacle blinds have been removed or turned to their correct operational position.
▷ Any management of change requirements for new processes and any modifications to existing processes have been met.

Apart from the safety critical elements of a facility, there is a need for some form of emergency safety system that will allow the process and production systems to shut down automatically if things go wrong.

This shutdown system will have various levels of command in order to react appropriately to various situations. For example, if smoke were detected in a boiler room, it is unlikely the whole process would be shut down in the first instance. A more appropriate level of command might be to shut down equipment in the affected area, sound an alarm to warn personnel of the smoke and to instigate an immediate investigation.

Within this system there will usually be:

▷ An Emergency Shutdown (ESD) system. This may shut down specific sections of the process or the full process, depending on the situation. Activation may be applied automatically or manually.
▷ Emergency shutdown valves – Isolation and High Integrity Pressure Protection System (HIPPS) – in order to isolate a specific section.
▷ Blow down (vent) valves and system in order to vent gases or vapour away safely.
▷ A fire and gas monitoring system. This will activate the Emergency Shutdown (ESD) system if fire or gas is detected.
▷ An automatic fire deluge system to contain any fire detected by the fire monitoring system.
▷ A temporary safe refuge for personnel in case of an incident.

9.4 Control of ignition sources during maintenance and operations

9.4.1 Sources of ignition

Potential sources of ignition come in many forms, and understanding where they might emanate from is the first step in controlling them. They include:

▷ Cigarettes and matches
▷ Heated process vessels, e.g. dryers and furnaces
▷ Flames from cutting and welding
▷ Lightning strikes
▷ Electric sparks from electric motors and switches
▷ Electromagnetic radiation of different wavelengths

9

▷ Electrostatic discharge sparks
▷ Impact sparks, friction heating or sparks
▷ Direct-fired space or process heating
▷ Vehicles (unless they have been specially designed or have been adapted)

All areas designated as hazardous need to have systems and control measures in place to control ignition sources. These control measures should include:

▷ The implementation of a permit-to-work system for maintenance activities which may generate sources of ignition, e.g. grinding, use of a blowtorch, etc.
▷ There should be strict working procedures in place during periods when hazardous activities are being conducted, e.g. during the loading and unloading of tankers.
▷ Prohibition of smoking and the use of matches or lighters.
▷ Only using electrical and mechanical equipment and instrumentation which has been designed and manufactured for use in the zones in which it is to be used. This includes the use of high intensity electromagnetic radiation sources (e.g. limitations on the power input to fibre optic equipment and avoiding the use of high intensity lasers or other sources of infrared radiation).
▷ Controlling the risk from pyrophoric scale (this is usually associated with the formation of ferrous sulphide within process equipment).
▷ The appropriate selection of vehicles and internal combustion engines for zoned areas in which they are to be used (this should include such equipment as spark arrestors on exhaust systems). There should also be controls over the use of normal vehicles.
▷ The earthing of all plant and equipment.
▷ All equipment should be electrically bonded so there is no difference in electrical potential between equipment which might result in static electrical spark.
▷ The installation of lightning protection, which includes having snuffing systems on vent stacks.
▷ The elimination of surfaces which are above the auto-ignition temperature of any flammable substances or materials which are being handled or stored.

Sources of ignition – direct-fired heaters, hot oil systems and processes operating above auto-ignition temperatures

There are a number of processes within the oil and gas industry which use direct-fired heaters. These heaters need a fuel supply for them to function, but if that fuel supply were to experience a leak close to the furnace, it would be reasonable to expect that any fuel forthcoming would find a source of ignition.

Consequently, maintaining a safe working environment in these situations should be achieved by a combination of a high standard of integrity of fuel and process pipelines, together with a means of rapid detection and isolation if any of the pipes do fail. Finally, there should be a means of rapidly isolating any pipework which may experience a fuel leak.

Where there is potential for pipework carrying process material to fail, this should be the subject of a Hazard and Operability (HAZOP) study, as the consequences can be regarded as very serious.

Sources of ignition – lightning

We will deal with lightning protection in more detail in Chapter 15.

Lightning protection systems are a specialist area for consideration. However, in general, the protection measures should include the following:

▷ All metal containers should be of sufficient thickness to withstand a current surge (lightning strike). This is usually a minimum of 5 mm.
▷ Air termination wires fitted above containers and storage tanks. These wires should be non-conductive electrically.
▷ Down-conductors fitted to all other metal structures. These down-conductors should be in sufficient numbers so as to subdivide any current surge (lightning strike) adequately.
▷ All earthing systems should be interconnected to a single earth termination system. This system usually takes the form of a mesh or grid pattern around the site.

Sources of ignition – vehicles

Most vehicles present a number of sources of ignition. These include the exhaust system of combustion engines, their electrical systems including ignition timing devices, hot brakes, potential electrostatic discharges from static build-up, etc. Local rules should govern where normal vehicles can be taken and where they should be excluded.

Many onshore facilities, especially refineries, will host the loading and unloading of tankers, including sea tankers, road tankers and rail tankers. These operations present their own hazards, including the release of gas or vapour. Consequently, in these circumstances controls will need to be implemented in order to control the release of gas or vapour, as well as controls over sources of ignition. These should include implementing safe systems of work.

We shall be dealing with this aspect in greater detail in Chapters 19 and 20.

There are many circumstances when process plant and equipment has to be depressurized, vented, gas freed and/or purged. Some of these occasions provide a window of opportunity to carry out specific maintenance activities which are only possible when the plant or equipment is in this condition, i.e. breaking containment, vessel entry for inspection, equipment replacement, etc.

Let's have a more detailed look at what these operations entail.

9.5.1 Gas freeing and purging operations

Gas freeing and purging involves filling spaces in plant or equipment that are currently filled with vapour, with an inert gas, sometimes known as a blanket gas.

This then prevents any flammable mixture forming due to the ingress of blanket gas into the system. This effectively enriches the atmosphere in the vapour space with inert gas, in order that the atmosphere be taken to a level below its lower flammable limit.

9.5.2 Inerting operations

Gas freeing and venting are used to control many types of processes and storage facilities. The partial or complete substitution of a flammable or explosive atmosphere in a contained environment by an inert gas is a very effective means of preventing an explosion. However, inerting is normally only undertaken when the hazard presented by the volatile atmosphere cannot be eliminated by other means such as the adjustment of process conditions to ensure substances are below flammable limits.

Inerting is typically used in storage tanks where a material may be above its flashpoint. Inert gases are also used in the transfer of flammable liquids under pressure, such as the transfer of hydrocarbons from seagoing oil tankers to land based facilities. We will be looking at this particular aspect in greater detail in Chapter 19.

To ensure the inerting system is not compromised by the ingress of air, which will reduce the ability of the system to present a non-flammable/non-explosive atmosphere, a slight positive pressure should be applied to the inerting system.

One of the major risks associated with use of inerting is that of asphyxiation, particularly in confined spaces. Where people have to enter a confined space, it should be a requirement that a formal control system is implemented, such as a permit-to-work system. This will ensure appropriate procedures and control measures are applied, such as replacing the blanket gas with fresh air and testing oxygen levels prior to entry.

Maintenance activities inside confined spaces such as storage tanks are fraught with hazards. Even if tanks have been cleaned, there is still the possibility of residual flammable material being present in cavities and joints. Ideally, tanks should be cleaned using detergents or steam. During maintenance activities inside tanks, a constant flow of fresh air should pass through the tanks in order to disperse any flammable vapour that may have been trapped, or is emanating from residual product missed in the cleaning operation.

9.5.3 Venting operations

During any process operation, there will always be a need to vent the system if an overpressure problem occurs, which will then activate a pressure relief valve. The venting system is made up of a series of pipes which are connected to each pressure relief valve. These pipes then channel the overpressure through to the vent stack where the overpressure is released to atmosphere in a safe and controlled manner (blown down). Alternatively, the overpressure can go to a flare where it is allowed to combust (burn) in a controlled manner.

Other drain and vent systems go via a closed drain header to a process vessel, known as a Gas Zone drum (GZ drum) in the collection system where the dissolved gases from the produced water are separated or flashed out of the liquid and sent to the flare. The residual liquid is pumped to a 'slop system' which stores, and ultimately recycles, the accumulated liquids into the process.

Finally, venting in the form of Pressure and Vacuum relief valves (P&Vs) are used on many storage tanks, process vessels and road and rail tankers when filling and emptying operations are ongoing. This allows pressure in the tank/vessel to be equalized with the atmospheric pressure, thus eliminating the risk of rupture caused by overpressure, or collapse caused by vacuum.

9.5.4 Drained water

All hydrocarbon processes produce a mixture of oils, gases, solids and water, all of which must be separated so they can be dealt with individually.

This separation process takes place in numerous pieces of equipment such as inlet separators, production separators, 3-phase separators, slug catchers, surge drums, filters, etc.

The processes use various methods in order to separate the water from the oil and gas. These include dehydration processes, pressure let down, level control, interface, filtration, demisting, etc.

9

Water which has been separated out will usually be drained into a pressurized tank and allowed to settle before being cleaned in an oily water separator. From there it will be discharged into the sea if it is offshore or into drainage systems if it is on land.

Legislation is likely to cover what can and cannot be discharged from processes offshore into the sea and onshore into land drains.

It is also worth mentioning that hydrate formation can occur when draining water from pressurized Liquefied Petroleum Gas (LPG) spheres. This operation should be carried out using a sequential valve procedure and only undertaken by trained and authorized personnel. A prime example of the catastrophic consequences of not adhering to sequential valve procedure was the 1966 Feyzin (France) BLEVE disaster referred to in Chapter 3, section 3.4, 'The concept of hazard realization'.

REVISION QUESTIONS FOR ELEMENT 2 CONTINUED

9   **Explain** what risk based maintenance is.
10 An operator is required to drain water from an LPG sphere.
    (a) **Identify** the main hazard associated with this operation.
    (b) **Identify** the control measures that can be applied in order to reduce this hazard.

You can compare your answers with our guidance at the end of the book under the section entitled 'Revision and assessment guide'

CHAPTER 10

Element 2

Sub-element 2.6:
Start-up and shutdown

This chapter covers the following learning outcome

Outline the hazards, risks and controls to ensure safe start-up and shut down of hydrocarbon containing equipment and processes.

10.1 Hazards and controls associated with safe start-up and shutdown

10.1.1 Safe start-up and shutdown

The most hazardous part of operating any plant or process system is the starting up or shutting down procedure of that system. This is because the system is constantly evolving until it reaches its optimum level, whether that is a level of normal operation, or a level of being fully shut down.

This process of evolvement is a constant changing of the energies within the system (increasing or decreasing). These energies can be pressure, temperature, flow, etc., and the plant or process system must be able to withstand these changes.

It is essential that all personnel involved in either of these operations work as a team from a predetermined operating procedure which, as well as being kept fully up to date, takes into account all potential eventualities associated with start-up or shutdown.

Safe start-up and shutdown – operating instructions and procedures

As we've just mentioned, it is essential that the operators have the 'operating procedure' available for reference at all times.

The 'operating procedure' document should take into consideration the following points:

- There should be no easier, more dangerous alternative procedures other than that specified in the operating procedure documentation.
- A review system should be in place to ensure that the operating procedures are kept up to date and also to detect any errors in procedures which can be quickly corrected.
- The personnel involved in the operation should be involved at the design stage of the operating procedures to ensure that the document does not become too prescriptive.
- Operating procedures should include information about the requirements necessary for the use of personal protective equipment when carrying out the operating procedures.
- At the start of the procedure the document should include:
  - ▷ an overview of the work to be done
  - ▷ any risks which the operator may be exposed to, based on a risk assessment of the work
  - ▷ any prerequisites should be clearly stated in order that a check can be made to ensure that it is safe to proceed with the procedures.
- Each operating procedure document should be dated and an expiry date indicated where appropriate, e.g. valid for six months from the date given on the document.
- The document should make it clear which procedures apply to which situations. There should be no ambiguity – an appropriate method of coding each procedure should be used.
- The document should:
  - ▷ ensure the most important information on the page is clearly identified and made the most prominent on the page
  - ▷ ensure different sub-tasks are listed under separate headings to differentiate them
  - ▷ be written in a language which is simple and familiar to the operators who will be performing the tasks
  - ▷ ensure that the nomenclature (terminology) used is consistent with that used on controls or panels
  - ▷ ensure that any warnings, cautions or notes are put immediately before the instruction step to which they refer
  - ▷ ensure that any shapes, symbols and colours which are used for graphics are consistent and conform to industry standards.

When a plant or process system is started up, it has the potential to present additional hazards over and above those associated with normal process operations. Some examples include:

- The mixing of air and hydrocarbons (potentially explosive mixture)
- The mixing of oil and water
- Temperatures below freezing with the potential for any water present in the system to freeze (hydrate formation within pipework and valves)
- Equipment potentially being subjected to over- or under-pressurization
- Equipment potentially being subjected to thermal or mechanical shock
- The introduction of corrosive and toxic fluids and gases
- The potential for hydrocarbon release (resulting in fire/explosion)
- The potential for mechanical equipment failure

Prior to the start-up of any process it is essential that all Safety Critical Elements (SCEs) are fully operational and can be depended upon to operate if a problem occurs. Depending on the particular process system, these SCEs might include:

- Fire and gas detection systems
- Emergency shutdown systems
- Vent and blow down system, high pressure and low pressure
- Fire deluge protection systems
- Emergency power supply systems
- Flare

Safe start-up and shutdown – thermal shock

Thermal shock is one of the greatest hazards to be considered when a plant or process is started up. This is because it has the potential to create a catastrophic event if a critical component is fractured or fails.

Although we shall be looking at 'thermal shock' in more detail in Chapter 11, we will now look at what is, and causes, this phenomenon.

Thermal shock is where a material is exposed to a sudden and significant change in temperature. This results in the material expanding at different rates within a limited area, causing a crack or failure. An everyday example of thermal shock is if boiling water is poured into a glass jar. As we all know, the glass jar breaks and this is because the molecular structure on the inside of the glass expands more rapidly than the outside of the glass, causing it to break. Under certain conditions, the same effect can be experienced on metal pipework.

A typical example from the oil and gas industry might be if superheated steam is introduced into pipework which has not been previously warmed; this may well cause thermal shock in the pipework, resulting in a fracture and an escape of superheated steam. Let's take a look at case study 10.1, which is a well documented incident involving thermal shock.

Case study 10.1

The Esso Longford gas explosion in Australia on 25 September 1998, which killed two people and injured a further eight, was a prime example of thermal shock causing a catastrophic event.

The investigation that followed the incident determined that a pump used for pumping hot oil into a heat exchanger went offline for four hours because of a complication in the process system, and taking it offline was part of normal procedures. However, during this down time it is estimated that the pump experienced temperatures as low as minus 48°C and ice had formed on the unit. It was decided therefore

that to thaw out the pump, pumping warm oil should resume. However, the temperature differential between the pump and the heat exchanger it serviced (−48°C) and the oil introduced into the system to warm things up (+230°C) was enough to cause a brittle fracture in the heat exchanger and allow hydrocarbon vapour to escape to atmosphere.

The resulting vapour cloud drifted to a nearby set of heaters and ignited. The flame front then worked its way back to the ruptured heat exchanger, which caused a jet fire which lasted for two days.

Thermal shock can be reduced by:

▶ The gradual introducing of steam or warm product from a lower temperature base (i.e. not superheated steam at the outset)
▶ Thoroughly warming up the system prior to use
▶ Designing in expansion loops into the system, for example bellow pieces. These allow the pipework to expand with thermal change without compromising the integrity of the system.
▶ Using materials with greater thermal conductivity
▶ Reducing the coefficient of expansion of the materials
▶ Increasing the strength of the materials

10.1.2 Example of a start-up operation

We will now look at an example of a start-up operation. This example involves a gas production train on a gas producing installation.

It is standard practice to ensure that only essential personnel are allowed into the production area whilst the start-up operation is ongoing. The lining out procedure is a joint operation carried out by both the control room operators (CROs) in the control room and the operators on the platform. Throughout the whole start-up operation, it is essential that close communications are maintained between the control room and the plant operators, usually by radio.

All Safety Critical Elements (SCEs) are checked to ensure they are operational. All pipes and drains opened during maintenance activities are closed off as required, all instrumentation has been replaced, calibrated and in good working order, and all spades, blinds or spectacle blinds have been removed or turned to their correct operational position.

The operating procedure document is checked to ensure it is up to date and that copies of the same document are possessed by all parties including those physically on the line as well as those in the remote control room.

Once all preliminary procedures are satisfactorily completed, the production train is walked and lined

out as per the pre-start-up procedure set out in the operating procedure document.

This will entail operators opening valves on the system in a specific sequence. However, as the sequence is critical, it is not just left up to the operator following the operating procedure document. In this case, each valve is fitted with a unique, specially designed key. Only when the correct valve is put into the correct position can the key be removed to open the next valve in the sequence. The key is designed so that it will only fit the correct valve in the sequence.

This process is repeated until all the valves are put into their correct position and in the correct sequence, resulting in a foolproof procedure for a highly critical operation. Once all the valves have been set, the final key taken from the last valve allows the operator to open the isolation valve. This then allows gas to be introduced to the system and the process becomes operational.

Let's look at what these valves look like in real life. The examples shown are with kind permission of ISS Safety Systems Ltd (Figures 10.1, 10.2 and 10.3).

10

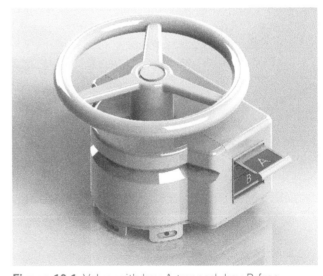

Figure 10.1 Valve with key A trapped, key B free
Source: ISS Safety Ltd.

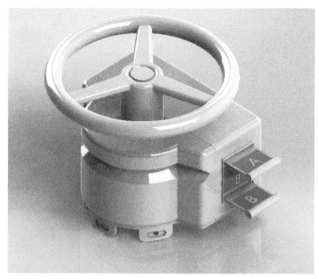

Figure 10.2 During the operation both keys are trapped
Source: ISS Safety Ltd.

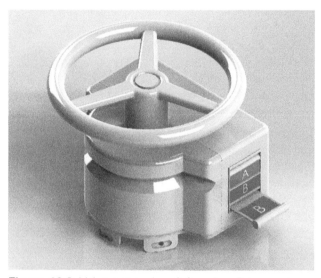

Figure 10.3 Valve open – key A free, key B trapped
Source: ISS Safety Ltd.

The control room will monitor the setting of the valves, and the CROs have the ultimate control of these valves, although they can be closed manually if required or by the Emergency Shutdown system (ESD) in the case of an emergency.

Once all the valves are set and the isolator valve is opened, gas from the well is introduced into the system. At this stage of the start-up operation, the gas pressure is set to a low level and only increased gradually as per the start-up procedure document. Constant checks will be made to the system to ensure nothing is amiss. This will include monitoring readings in the control room as well as watching and listening for leaks by the operators at the plant.

This is the stage where hydrates (ice plugs) might form, and as a counter measure, methanol may be injected upstream.

Only when all optimum pressure levels, temperatures and flows have been established, can the line be regarded as being 'online'. Furthermore, it is only when this level of operation is attained that the control room can take over the operations and run the operation automatically.

10.1.3 Shutdown

Shutting down a process can be equally as hazardous as start-up. Consequently, the personnel involved should work as a team from a predetermined operating procedure which will take into account all possible eventualities associated with shutdown.

In the case of our example of a gas production train on a gas producing installation, the procedure is undertaken in a deliberately slow manner. Each well (if there is more than one) is taken offline by closing its wing valve.

Once the gas flow has stopped, all the ancillary systems can be slowly shut down and isolated.

The production train is shut down and isolated by closing the valves in the reverse order to that in which they were opened. Again, the key system on each valve ensures the correct sequence is followed.

Looking at another example of shutdown, this time on a refinery, the same sources of hazard exist. In this case, the procedure will be undertaken in a number of phases, including the following:

▶ Cooling and depressurization of the process
▶ Reducing the flow and process liquid levels
▶ Removing any residual hydrocarbons
▶ Inerting and gas freeing the system
▶ Removing any corrosive and/or toxic materials from the system
▶ Removing any water from the system
▶ Blinding operations (the isolating of process pipework)

Over and above that, any fired heaters should be shut down, reducing the heat input into the process. The flow of product should be gradually reduced and eventually shut off. Any excessive pressure emanating from hydrocarbon gases should be released into the gas collecting system.

10.2 Hazards and controls associated with water and hydrates, their presence and removal

10.2.1 Water and hydrates – their presence and removal

The formation of hydrates (ice plugs) within a production process plant during start-up and shutdown was mentioned as a potential problem earlier in this chapter. Because this problem has the potential for serious consequences to evolve, it deserves to be discussed in greater depth.

A hydrate is formed when water is trapped within the oil or gas process and the conditions within the process allow it to freeze into a solid plug.

When the flow of product or the pressure of the product is reduced, this has the effect of reducing the temperature and it is at this point where the formation of a hydrate can occur if there is water within the system.

Hydrates can also be formed in pipelines which run across the sea bed. This is because temperatures at sea bed level can be low enough to freeze un-salted water which accompanies the oil and gas as it is extracted from the wells.

Some hydrates are made up of additional materials, as well as water, and these can form into sticky crystals which have the potential to grow into large ice plugs which can completely block a pipeline.

10.2.2 Water and hydrates – controls

To prevent hydrates forming in the first place, antifreeze (usually methanol or glycol) can be introduced into the process system so that it forms part of the product. This antifreeze is usually recovered later when the oil and gas arrives onshore, allowing the antifreeze to be reused.

However, using antifreeze is an expensive process, and an alternative means of control is the use of trace heaters along with insulation. Trace heaters are cables which run along the outside of the pipework and which apply warmth to the system. The escape of this warmth is reduced by the insulation applied to the pipe.

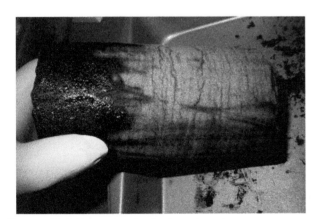

Figure 10.4 Example of a hydrate ice plug
Source: US Department of Energy.

10.2.3 Removal of water

The removal of water from processed oil and gas can be achieved by a number of methods. Each has its own advantages and disadvantages.

The following techniques and technologies outline what means are available for removing water from oil and

gas. We have also indicated their relative advantages and disadvantages.

Removal of water – gravity separation

Where the product is a mixture of oil and water and is allowed to stand for a while, water will naturally settle to the bottom, as the specific gravity of water is greater than oil. This can then be drained off.

The effectiveness of this system is dependent upon how long the system was allowed to stand before water is drained off, and whether the temperature is low enough to reduce the saturation point to an acceptable level. Lowering the saturation point ensures as much water as possible leaves the system.

Some types of oil have the characteristic of holding water in suspension rather than allowing it to settle at the bottom and separate out. In these cases, gravity separation is a less than effective option.

The main advantage of this system is the relatively low cost of operation.

The main disadvantage is that it only removes free water, and any emulsified water will remain in the product.

Removal of water – centrifuge (spin the oil clean)

This system uses the principle of centrifugal force and the fact that water has a different specific gravity from oil, and that difference allows this system to separate the two materials. The greater the difference in specific gravity between the oil and the water, the more effective the process is. Consequently, centrifuges work best on low specific gravity and low viscosity oils such as turbine oils, rather than other, heavier oils.

Centrifuge systems remove both free water and emulsified water, but their effectiveness will depend on how low the specific gravity of the oil is when the centrifuge process takes place.

The main advantage of this system is that it can, in the right circumstances, remove both free and emulsified water. It can also remove other heavier contaminants and has a comparatively high throughput compared with other technologies. Centrifugal separators are relatively cost-effective.

The main disadvantage of this system is that the capital cost is relatively high. Also, cleaning a centrifuge is a time-consuming task which has to be undertaken frequently.

Removal of water – absorption removal

Absorption removal is the process of using filters to absorb the moisture from product as the product passes through the filter. There are a number of different types of filter medium. Cellulose based filters are effective for removing small amounts of water. Also, as they swell

10

quite noticeably when they do absorb moisture, they are a useful means of indicating the presence of water in the product.

Other types of filters have an outer wrap consisting of polymer and desiccants which can absorb both free water and emulsified water. They can also filter out solids.

The main advantage of absorption removal is that it is a relatively cost-effective means of removing small amounts of water.

The main disadvantage is that each filter has a limited capacity.

Removal of water – vacuum dehydration

The process of vacuum dehydration works on the principle that water boils at a lower temperature when it is at a lower pressure. Consequently, at 0.9 bar, water will boil at around 52°C.

A vacuum dehydration unit reduces the pressure of the product within it. Air which has been dried and warmed is then passed over the product and the moisture is then transferred to the air from the product in the form of steam vapour.

The efficiency of the system is enhanced by increasing the surface area of the product as it is exposed to the warm air. This can be achieved by exposing the product in shallow amounts over numerous surfaces within the dehydrator.

The main advantage of this system is its ability to remove moisture to very low levels. It also has the ability to remove other impurities such as fuels and solvents.

The main disadvantages are its cost and relatively low flow rates.

Removal of water – air stripping

Air stripping is another form of vacuum dehydrator which works by mixing air, or nitrogen gas, with a stream of heated product within the air stripping unit. The gas then absorbs the moisture from the product and when the gas and product mixture is expanded, the gas separates from the product and takes the moisture with it.

The main advantage of this system is that it costs less to maintain than a typical vacuum dehydrator as it has fewer moving parts.

The main disadvantages are its cost and relatively low flow rates.

Removal of water – heating the oil dry

Some processes, because they run at elevated temperatures, are self-cleansing due to the fact that water naturally evaporates at these temperatures. However, these processes must be controlled effectively to avoid harm, particularly with mineral oils.

Compared to centrifugal and vacuum separation systems, heating the oil dry can be an effective means of removing water from a product.

Summary of methods of hydrate removal

The final decision as to which system of water removal is appropriate to any one process will depend on:

1　The volume of product to be treated
2　The required level of moisture content in the final product

If the level of moisture required in the final product is low but the saturation levels of the product prior to treatment are high, then more complex and expensive systems will be required.

10.3 Hazards and controls associated with testing, commissioning and hook up

10.3.1 Commissioning

The commissioning of a process plant involves undertaking tests on the plant prior to it going into production in order to determine that it will function adequately and safely. The commissioning process also includes training the people who will operate the plant as well as the control room staff who will oversee operations. Finally, it includes writing the operating procedure document.

During this commissioning period, when operating experience is at a minimum, the possibility of unforeseen events occurring cannot be eliminated. Consequently, all safety precautions should be reviewed prior to commissioning commencing. This is normally undertaken at the design stage of the plant and usually takes the form of a Hazard and Operability (HAZOP) study or an equally extensive and in-depth risk assessment.

The process of commissioning involves taking the plant through a number of phases, all of which have to be completed in a satisfactory manner before the plant can be handed over as fully functional and ready for operational use.

A typical sequence of phases undertaken during the commissioning process is set out below.

1　The system configuration is checked (walking the line).
2　The pipework and system integrity is checked.
3　The instrumentation system is checked.
4　All alarms are verified as working.
5　All lines and vessels are flushed and cleaned.
6　All ancillary equipment is inspected and assessed as to its adequacy.
7　All instruments and vessels are calibrated.
8　The start-up protocol is established.

9 The shutdown protocol is established.
10 Commissioning trials are undertaken.
11 The plant is hooked up.
12 The plant is handed over.

We will now take a look at each of these phases in a little more detail.

Commissioning – system configuration check (walking the line)

This phase is where all the pipework and connections are physically inspected to ensure their configuration is as it should be and they have been installed correctly. This inspection is made against Engineering Line Diagrams (ELDs) also known as 'as built' drawings. The inspection process also includes ensuring the plant and equipment are clean, empty and fit for purpose.

Commissioning – pipework and system integrity

This phase is where all the parts of the system are pressure tested to ensure there are no leaks or unexpected deformities. This is usually undertaken using hydrostatic testing techniques.

Hydrostatic testing is a system which tests the integrity of plant and equipment, including pipework, which will be subject to pressure under normal working conditions. It involves filling individual sections or components of the system with a liquid, usually dyed water, and then pressurizing that section or component. Pressure is then monitored to see if there is any unexpected deformity or loss of pressure. A loss of pressure would indicate a leak, and the fact that the water is dyed will help locate any leak.

Once all the sections and components have been individually tested, all the sections and components are connected and a test of all the flanged joints and couplings is carried out. This is carried out using nitrogen and a trace gas in order to detect any leaks.

Commissioning – instrumentation system check

This phase is where the system is checked against the Piping and Instrumentation Diagrams (P&IDs). A P&ID is a schematic drawing showing the physical sequence of equipment and systems, as well as how these systems connect. Included in the P&IDs are:

▶ All the instruments, their location and designation
▶ All the mechanical equipment with their names, numbers and functions
▶ All the valves with their identification and function
▶ All the piping with their sizes and identification
▶ Any miscellaneous equipment such as vents, drains, sampling lines, etc.
▶ Start-up and flush lines
▶ The direction of flow

▶ Any interconnection references
▶ All control inputs, outputs and interlocks
▶ All computer control system inputs
▶ Any identification references of components and sub-systems delivered by third parties

Commissioning – verification of alarms

This phase is where all the alarm settings, microprocessor signals, hardware trips and all other instrumentation are checked to ensure they are set correctly and are functioning as required. The computer systems are also checked to ensure they are functioning properly and that information from all the field instruments is being relayed to the computer interfaces and is being displayed correctly.

Commissioning – flushing and cleaning of lines and vessels

This phase is where all the vessels and associated pipework are flushed through with water in order to flush away any contaminant material. The process is also aimed at ensuring there are no foreign items or materials within the system which may have been left behind at the time of construction.

Once the flushing operation has been completed, it may be necessary to dry the system completely, especially if the final process system is incompatible with water. This is done by venting warm, dry air through the system so that any residual moisture is evaporated away.

Commissioning – inspection and assessment of ancillary equipment

This phase is where all the ancillary equipment, such as pumps, fans, heat exchangers, condensers, compressors, etc. are tested to ensure they perform to their expected requirements.

Commissioning – calibration of vessels and instruments

This phase is where all the vessels and instruments are checked to ensure they have been calibrated correctly.

Commissioning – start-up protocol

This phase is where the procedure for starting up the installation from complete shutdown to full production is evolved and set down in writing. This phase also includes establishing all potential erroneous occurrences and providing guidance on the best and safest course of action on how to deal with each of them. This start-up procedure will form part of the 'operating procedure' document.

This phase will also involve the training of all personnel involved in the operation of the system once it is handed over.

10

Commissioning – shutdown protocol

This phase is where the procedure for shutting down the installation from being in full production to complete shutdown is evolved and set down in writing. This phase also includes establishing all potential erroneous occurrences and providing guidance on the best and safest course of action on how to deal with each of them. This shutdown procedure will form part of the 'operating procedure' document.

This phase will also involve the training of all personnel involved in the operation of the system once it is handed over.

Commissioning – commissioning trials

This phase is where the plant is started up, run for a predetermined length of time within specific parameters, and then shut down. This is so the plant can be proven to be operational. It is also an ideal opportunity to give some 'hands on' training to the personnel who will be operating the plant once it is handed over.

Commissioning – hook up

This is the phase where the system is connected to – although isolated from until the initial start-up procedure is ready to be undertaken – its source of product (oil, gas, etc.). It also includes the system being connected to – but again being isolated from – the utilities needed to operate the system.

Commissioning – handover

This is the final phase of the commissioning process and involves handing over the responsibility for the plant, as well as the plant itself, to the operational department. All associated documents, drawings and procedures will also be handed over.

During the premier run of the plant, it is quite normal for representatives of the suppliers of the plant and/or equipment to be present or on hand.

REVISION QUESTIONS FOR ELEMENT 2 CONTINUED

11 Thermal shock is regarded as one of the most significant hazards to be considered when a plant or process is started.
    (a) **Explain** what thermal shock is.
    (b) **Explain** how the effects of thermal shock can be reduced.
12 There are around twelve stages in the process of commissioning a process plant or system. **Identify SIX** of those stages.

You can compare your answers with our guidance at the end of the book under the section entitled 'Revision and assessment guide'.

Sources of reference for Element 2

American Petroleum Institute. Corrosion management information. Available at: www.api.org

British Standards Institution (2001) *BSEN ISO 15544 Petroleum and Natural Gas Industries – Offshore Production Installations – Requirements and Guidelines for Emergency Response*. London: BSI. ISBN 978-0-80-70983-8.

Health and Safety Executive (n.d.a) 'Human Factors: Safety Critical Communications'. HSE. Available at: www.hse.gov.uk/humanfactors/topics/common3.pdf

Health and Safety Executive (n.d.b) 'Maintenance of Work Equipment'. HSE. Available at: www.hse.gov.uk/work-equipment-machinery/maintenance.htm

Health and Safety Executive (n.d.c) 'Plant Modification/Change Procedures'. HSE. Available at: www.hse.gov.uk/comah/sragtech/techmeasplantmod.htm

Health and Safety Executive (1985) 'The Cleaning and Gas Freeing of Tanks Containing Flammable Residues'. Guidance Note CS 15. London: HSE. ISBN 978-0-7176-1365.

Health and Safety Executive (2005) *Guidance on Permit-to-work Systems. A Guide for the Petroleum, Chemical and Allied Industries* (HSG250). London: HSE. ISBN 978-0-7176-2943-5.

Health and Safety Executive Offshore Safety Inspectors (2006) *Improving Communication at Shift Handover*. London: HSE Books. available at: www.hse.gov.uk/humanfactors/topics/shifthandover.pdf

Health and Safety Executive (2009) *Safety and Environmental Standards for Fuel Storage Sites. Process Safety Leadership Group Final Report*. London: HSE Books. ISBN 978-0-7176-6386-6. Available at: www.hse.gov.uk/comah/buncefield/fuel-storage-sites.pdf

Health and Safety Executive (2011) *Managing Contractors. A Guide for Employers* (HSG159). 2nd edn. London: HSE Books. ISBN 978-0-7176-6436-8. Available at: www.hse.gov.uk/pubns/books/hsg159.html

Health and Safety Executive Hazardous Installations Directorate, Offshore Division (2007) *Key Programme 3 Report: Asset Integrity Programme*. Available at: www.hse.gov.uk/offshore/kp3.pdf.

Henderson, K. W. and Brazier, A. (2002) *Human Factors Aspects of Remote Operation in Process Plants* (CRR 432). London: HSE Books. ISBN-7176-2355-6.

Institution of Chemical Engineers (2006) *Safe Ups and Downs for Process Units*. 2nd revised edn. London: IChemE. ISBN 978-0-8529-5502-4.

Kletz, T. A. (1998) *What Went Wrong? Case Histories of Process Plant Disasters*. Oxford: Gulf. ISBN 0-88415-920-5.

Kletz, T. A. (2003) *Still Going Wrong: Case Histories of Process Plant Disasters and How They Could Have Been Avoided*. Oxford: Gulf. ISBN 0-75067-709-0.

Lardner, R. for Health and Safety Executive (1996) *Offshore Technology Report 0 OTO 96 003. Effective Shift Handover: A Literature Review.* London: HSE Books. Available at: www.hse.gov.uk/research/otopdf/1996/oto96003.pdf

Louca, L. A. and Boh, J. W. (2004) *Analysis and Design of Profiled Blast Walls. Research Report 146.* London: HSE Books. ISBN 0-7176-2808-6. Available at: www.hse.gov.uk/research/rrpdf/rr146.pdf

Mannan, S. (ed.) (2005) *Lees' Loss Prevention in the Process Industries: Hazard Identification, Assessment and Control, Third ed.* Amsterdam: Elsevier. ISBN 978-0-7506-7555-0.

University of Toronto (2012) Lockout/Tagout Standard. Available at: www.ehs.utoronto.ca/Assets/ehs+Digital+Assets/ehs3/documents/Lockout+Std+Feb+2012.pdf

U.S. Department of Labor Occupational Safety and Health Administration (1994) *OSHA 3133 Process Safety Management Guidelines for Compliance.* Washington, DC: OSHA Books. Available at: https://www.osha.gov/Publications/osha3133.pdf

U.S. Department of Labor Occupational Safety and Health Administration (2000) *OSHA 3132 Process Safety Management of Highly Hazardous Chemicals (Standards – 29 CFR 1910.119).* Washington, DC: OSHA Books. Available at: https://www.osha.gov/Publications/osha3132.pdf

10

CHAPTER 11

Element 3

Sub-element 3.1:
Failure modes

This chapter covers the following learning outcome

Outline types of failure modes that may lead to loss of containment from hydrocarbons.

Chapter contents

11.1 Introduction and terminology

11.1.1 Introduction

There is a wide range of materials used in the construction of facilities in the oil and gas process industry, the majority of which are steels of varying grades and types, although plastics are also extensively used, especially for the storage of hydrocarbons. These facilities include offshore rigs, onshore refineries and storage facilities.

The properties required of materials to cope with the wide variation of conditions can be extensive, and the process of considering which material is best for any given situation is a complex one. Consequently, understanding how certain conditions will affect any material in a given situation and then applying the most appropriate material to that situation is the ideal approach.

The first step in this approach is to understand the different ways in which the failure of materials can occur and what controls can be put in place to counteract these failures.

First of all we will look at some of the terms and phrases used in relation to forces to which materials can be subject and what they mean.

11.1.2 Tension

Tension – sometimes referred to as tensile force – is a force which is acting in two opposite directions. It could also be described as trying to stretch the material or pull the material apart. An example would be a chain being used by a crane to lift a heavy load.

All materials have a breaking point, and if we take our example and say the load lifted by the crane was greater than the threshold of the material the chain was made from, it will fail (break).

Figure 11.1 Tension
Source: Wise Global Training.

11.1.3 Compression

Compression force is the opposite of tension in that the material is being compressed by pressure. An example might be the legs of an offshore rig mounted on the seabed. The legs carry the weight of the rig and transfer this weight to its standpoint on the seabed.

Most materials will compress to some extent when pressure is applied. However, if the pressure exceeds the characteristic threshold of the material, given its current dimensions, the forces will have to be

dissipated in some way. Again, using our example of legs on an offshore rig, if the rig was exceptionally loaded, the pressure would likely be dissipated by the legs buckling.

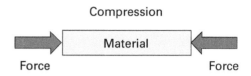

Figure 11.2 Compression
Source: Wise Global Training.

11.1.4 Tension and compression

We have covered tension and compression but under certain conditions, these two forces can be applied to one piece of material at the same time. For example, if a bending motion is exerted on a steel bar, this creates both tension and compression forces within the bar. A practical example of this is a steel beam suspended between two points which is carrying the weight of a piece of machinery.

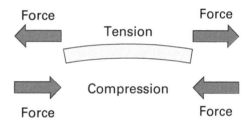

Figure 11.3 Tension and compression
Source: Wise Global Training.

11.1.5 Shear

Shear is where two forces are being applied to a material but in opposite directions (see diagram). The shear force acts on the dotted line in our diagram as if it is trying to tear the material apart.

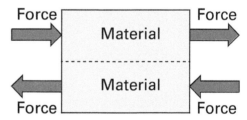

Figure 11.4 Shear
Source: Wise Global Training.

11.1.6 Properties of materials

Now we've looked at some forces which materials can be subject to, we should now look at properties of materials. Each material has its own properties, and understanding what these properties are will help us understand which materials are best applied to specific situations.

Properties of materials – ductile

A material which is ductile is one which can be subjected to tensile forces without it fracturing. An example of this is where a wire is drawn through an extrusion die.

Properties of materials – malleable

A material which is malleable is one which can have its shape changed without it cracking. This could be through compression (hammering) or by bending. Lead, tin and copper are all malleable materials.

Properties of materials – brittleness

A material which is brittle is one which has no plastic deformation characteristics. This means that if a force is applied which exceeds its characteristic threshold, it will crack or shatter with very little deformation. If the broken pieces were then brought back together, they would fit almost perfectly, much like a jigsaw. Cast iron is an example of a brittle material.

Properties of materials – elasticity

A material which has elasticity is one which can resume its former shape or dimension after a deforming force is applied and then released.

All materials, if they have the characteristic of elasticity, have a limit to how much force can be applied before permanent deformation enters the equation. If this limit is exceeded, the molecular structure of the material changes and the new shape becomes a permanent feature.

11.2 Failure modes

11.2.1 Introduction

Now we have had an introduction to some of the forces materials can be subjected to and some of the properties of materials, we can turn our attention to various failure modes experienced by materials used in the oil and gas industry.

The failure modes we shall be looking at are:

▶ Creep
▶ Stress
▶ Stress corrosion cracking
▶ Thermal shock
▶ Brittle fracture

11.2.2 Creep

Creep is where a solid material is subject to long term exposure of high stress levels and gradually deforms in shape or dimension. Creep is exacerbated when the material is also subjected to heat for long periods of time. Most cases of creep are insignificant but in certain situations, the amount of deformation may be unacceptable and may lead to failure of the component.

Control and monitoring of creep

The main control measure in managing or preventing creep is applied at the design stage of any facility. Process design should specify the use of appropriate materials in all situations which have the tolerances adequate for the expected parameters to which they will be exposed.

At the stage where the materials are in use within the process or application for which they were acquired, these parameters should be made known to the operators and should not be exceeded so that creep does not occur, or is kept within acceptable limits. Over and above these control measures, regular maintenance checks and inspections should be conducted.

If the processes are complex and/or the consequences of failure due to creep are significant, then more in-depth non-destructive checks may be appropriate, such as ultrasonic methods or the continuous monitoring of vibration.

There are certain situations where creep is a well-known risk factor, such as water tubes in boilers and furnaces. In these cases a creep failure would present itself as a rupture of the water tube.

In these cases monitoring and controlling tube metal temperatures is the ideal control measure. However, where super-heater and reheat super-heater water tubes are used, creep is a factor that can only be controlled by scheduled replacements of the tubes within their predicted life expectancy.

11.2.3 Stress

Stress is created when a load is applied to a material. For example, a girder carrying the weight of a piece of machinery is under stress due to the weight of the machine. Where heavy weights or loads are involved, there may be some visible deflection or movement, but this may be part of the design.

Stress can be described as 'the load divided by the cross-section area (newtons $\div$ m²)'.

Stress can occur during normal operations. For example, where components are subjected to warming and/or cooling, this will lead to those components expanding or contracting. It is essential that these movements and variations in sizes of these components are allowed for at the design stage, otherwise stress can be created at critical points such as anchor bolts and welds.

Stress is often built into components inadvertently during the manufacturing process, especially when welding has been part of that process or where components are bolted together, and especially when they have been forced into position.

On pipework which has been welded, stress is often unavoidably introduced due to the application of heat during the welding process.

However, these stresses can be mitigated to some extent by allowing the pipework, once it has been welded, to remain at a controlled temperature for a specific period of time by wrapping ceramic blankets around the pipework and then gradually reducing the temperature in a controlled way. This allows the pipework to find its natural configuration, which has the effect of reducing the stress levels introduced by the welding process.

This method of introducing measures that allow stress to dissipate naturally after components have been welded together is called Postweld Heat Treatment (PWHT). It is generally applied where the standards or codes of construction demand that inbuilt stresses must be reduced to an acceptable level, for example as in pressure vessels.

Postweld heat treatment involves heating the steel, immediately after welding has been completed and before it starts to cool, to a temperature dictated by the particular type of steel used, but this is generally around 230°C. This temperature is held for a period of time dictated by the thickness of the material, but in line with 1 hour for every 25 mm of thickness. The steel is then allowed to cool gradually under controlled conditions.

There is another, similar stress reducing treatment, called stress relief heat treatment, which is used on larger structures where the stresses have been built in as the structure has been constructed, and where the stresses are widely dispersed throughout the structure.

In this case, the whole structure is heated to a temperature dictated by the particular type of steel used, but it's generally around 650°C. This temperature is again held for a period of time dictated by the thickness of the material, but in line with 1 hour for every 25 mm of thickness. The steel is then allowed to cool gradually under controlled conditions.

The effect of both these treatments is to introduce conditions to the component or structure where residual stresses are allowed to relax and dissipate.

11.2.4 Stress corrosion cracking

Stress Corrosion Cracking (SCC) is where a material is subjected to both stress and corrosion. The corrosion has the effect of reducing the threshold of the material at a particular point (i.e. it weakens it), resulting in the stress causing the material to crack at that point. If the material had been subject to stress without corrosion, the crack would not have occurred. If the material had been subject to corrosion without stress, again, the crack would not have occurred. It is where these two factors combine that stress corrosion cracking can occur.

The environment can play a big part in introducing corrosion. For example, offshore facilities are subject to salt water as well as severe weather conditions. Cleaning can introduce chemicals which can be corrosive. Carbon steel pipework, which is lagged, may be subject to sweating.

Typical stress corrosion cracking failures occur in pressure vessels, pipework, highly stressed components and in systems when a deviation from normal operating conditions or the environment occurs.

SCC can be highly chemical-specific and some alloys can undergo SCC after a few exposures to a specific chemical environment. The surface of the material can appear bright and shiny whilst inside the material many microscopic cracks may be present. Consequently, SCC can lead to sudden and unexpected failure with possible catastrophic consequences.

Because stress corrosion cracking can be so highly chemical-specific, certain pairings of materials with specific environments should be avoided. These include:

▶ Brass paired with ammonia
▶ Stainless steel paired with chlorides
▶ High strength steel paired with hydrogen

The following controls will help in the prevention of stress corrosion cracking:

▶ Selection of appropriate material
▶ Controlling service stress
▶ Use of corrosion inhibitors
▶ Coating material
▶ Isolating material from the local environment

11.2.5 Thermal shock

Thermal shock is the stress introduced into a material as a result of a sudden and dramatic change in temperature. The result is that the thermal gradient of the affected area of the material is uneven, causing stress within the material. The temperature change causes the molecular structure of the material to expand, but as this expansion is not uniform, the bonds which hold the molecules in formation are weakened, which may lead to a failure (crack) in the material.

An everyday example of thermal shock would be if boiling water is poured into a glass jar. Inevitably, the glass jar breaks, and this is because the molecular structure on the inside of the glass expands more rapidly than that on the outside of the glass.

Let's look at an example of where thermal shock might occur. Superheated steam introduced into cold pipework would cause thermal shock on the pipework by expanding the inner surface of the pipes rapidly, whilst the outer surface would expand more slowly. This would create the possibility of cracks occurring in the pipe material when the strength of the material is overcome by the stresses of differing expansion rates. Furthermore, flanged joints may start to spring because the sections of pipework are expanding at different rates, and leaks may occur quite suddenly.

Expansion in pipes can be expected under normal operating conditions as well as exceptional circumstances such as the example just quoted. In order to prevent the critical effects of expansion on steam pipework, it is normal to design in expansion loops in order to allow the pipes to expand without building up stresses and strains at flanged joints. You can clearly see these loops (the U-shaped configuration) in the pipes on the right of the photograph in Figure 11.5.

Figure 11.5 U-shaped configuration in pipework
Source: iStock.

Expansion bellows are also used to prevent the critical effects of expansion. These are fitted directly into the pipework allowing the bellows to expand and contract, thus reducing the effects of expansion on the pipework and material.

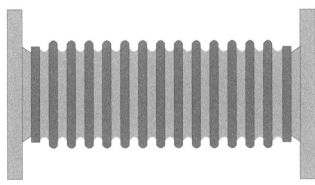

Figure 11.6 Expansion bellows
Source: Wise Global Training.

Other means of reducing thermal shock would be to:
▶ Change the temperature more slowly
▶ Change the temperature more evenly
▶ Increase the thermal conductivity of the material
▶ Reduce the material's coefficient of thermal expansion
▶ Increase the material's strength

11.2.6 Brittle fracture

An example of brittle fracture would be if someone tried to bend an engineering file, which is made of a very hard but brittle metal, it would snap. There would be little or no permanent deformation and the broken pieces of the file could be offered together in near-perfect alignment.

However, under certain circumstances low carbon and medium carbon steels, which are used widely in the petrochemical industry and which are normally considered to have ductile properties, can also be subject to brittle fracture.

The conditions where low and medium carbon steel can be affected by brittle fracture are as follows:
▶ A stress concentration must already be present in the steel. This may be a welding defect, a fatigue crack, a stress corrosion crack or a designed notch such as a sharp corner, a threaded hole or something similar.
▶ A tensile stress must be present. The tensile stress must be of a magnitude sufficient to cause deformation.
▶ The temperature must be relatively low for the steel concerned.
▶ Susceptible steel has been used.

The absence of any one of the above conditions will reduce the possibility of brittle fracture occurring.

Brittle fracture in normally ductile steel has caused many unexpected failures in the past, with some resulting in catastrophic consequences. Part of the reason for this behaviour is that it is not always necessary to have high stresses present for failure to occur. Brittle fracture can occur solely because of residual tensile stresses.

Brittle fracture of normally ductile steels occurs primarily in box-like structures such as box beams, tanks and pipes, particularly where their construction has involved welding.

11.3 What is meant by a safe operating envelope

A safe operating envelope is defined as the parameters and conditions a plant must operate within to ensure it is not subjected to excessive stress which might introduce or encourage failure modes. These parameters and conditions will be clearly set out in the operating procedure document. Any changes and adjustments to a process must remain within this safe operating envelope in accordance with normal operating procedures.

Some examples of parameters and conditions include:
▶ Safe working loads
▶ Maximum flow rates
▶ Maximum pressures
▶ Maximum and/or minimum temperatures

There are conditions, usually external environmental conditions, where a process has the potential to be operated outside of its safe working envelope. These

11

include high wind speeds, sea state, lightning, etc. In these cases the process may have to be shut down until conditions return to normal.

An example of a situation where parameters can be exceeded by adverse factors is as follows. A gas compressor unit is compressing gas in order to push it through a pipeline. Whilst the gas is moving forward, the equipment is within its safe working envelope. However, a discharge valve fails shut and the gas compressor unit goes into a condition known as 'surge'. This is where the compressor unit cannot move the gas forward through the outflow side so it tries to pump the gas through itself to the inflow side. The gas compressor unit is now outside its safe working envelope and is likely to destroy itself, with potential catastrophic consequences.

In order to counteract this particular scenario, it is necessary to anticipate this kind of event at the design stage, usually at the Hazard and Operability study (HAZOP) stage of development. In this particular scenario it would be possible to design into the process anti-surge valves and a bypass system which would allow the compressor to recirculate the gas without adverse effects until an operator could reset the process.

11.4 Use of knowledge of failure modes in initial design, process and safe operating procedure

Previously, we have discussed how failure can be introduced at the design and/or the manufacturing stage of a process plant. We have also discussed different failure modes, and finally we have discussed what the consequences of operating outside a safe operating envelope are.

Consequently, it can be seen that because these elements of failure may well lie unseen within the process plant, it is important that when the plant is in operation, it remains within the parameters of the plant that were set at the design stage.

At the very outset it is possible to ensure parameters are designed into the plant, which give the plant safe margins of operational limits.

When plant is being designed, careful thought and consideration should be given to the process itself and the extent of stresses that may be encountered in extreme circumstances and how these stresses might be controlled.

Some simple control techniques include:

▶ Fitting expansion loops and expansion bellows in steam pipes. We showed examples of these earlier.
▶ Ensuring the materials used are correct and capable of coping with the conditions they will be exposed to.
▶ Ensuring the thicknesses of the materials used are adequate to cope with the stresses they will be exposed to.

▶ Ensuring the strength of materials used is adequate.
▶ Ensuring equipment is correctly supported.
▶ Ensuring pipework and machinery is correctly aligned.
▶ Automatic shutdown trips are set to activate where parameters are exceeded.
▶ Ensuring the control systems are adequate.
▶ Having operating procedures to hand.

11.5 Failure of the annular rim (bottom rim of storage tank)

The bottom plate in a storage tank is known as an annular plate and the junction between this bottom plate and the wall of the storage tank is known as the annular rim. The annular plate usually sits on a foundation of hardcore or a concrete ring wall, and is joined to the walls of the tank. Although the annular plate is not subject to high levels of stress, the joint of the plate to the walls is. This is because the weight of the product within the tank wants to push the walls outwards whilst at the same time push the annular plate downwards. This creates a high level of bending stress. The quality of the foundations will have a bearing on the downward deflection of the annular plate.

A further complicating factor to this stress is the fact that annular plates are prone to corrosion attacks both on the outer side where the tank shell sits on the annular rim and on the underside of the annular plate where trapped water may lie undetected. This corrosion, coupled with the prolonged stress, can lead to stress corrosion cracking and failure occurring without warning.

Control measures should include ensuring the annular plate is kept as dry as possible by keeping the bunded area, where the tank sits, well drained. We shall be dealing with methods of draining bunded areas in Chapter 14, in the section on bunding of storage tanks.

Other control measures include regular inspections of the annular rim and the application of effective maintenance.

REVISION QUESTIONS FOR ELEMENT 3

1 **Explain** what stress corrosion cracking is and under what circumstances it might be encountered.
2 **Define** what a safe working envelope is.
3 Many fixed and floating roof storage tanks have annular rims.
   (a) **Outline** what an annular rim is.
   (b) **Explain** what problems can be associated with annular rims.

You can compare your answers with our guidance at the end of the book under the section entitled 'Revision and assessment guide'.

Sub-element 3.2:
Other types of failures

12.1 Weld failures – the need for regular weld inspection and non-destructive inspection techniques

12.1.1 Introduction to weld failures

When steel structures are fabricated, the most common method of joining the steel components together is by welding. Consequently, the need for welds that are of good quality and properly applied is of great importance.

There are many types of welding processes, including:

▶ Electrode (stick)
▶ Manual Inert Gas (MIG)
▶ Tungsten Inert Gas (TIG)
▶ Brazing techniques using oxyacetylene gas

The reasons why welding is a favoured method of joining steel components together is because it is speedy and efficient and results in a joint which has strength and is usually reliable, although there are exceptions to this reliability factor which we shall discuss later in this chapter.

When a weld is being formed, certain chemical and metallurgical actions take place in the metal adjacent to where the weld is being deposited. Understanding what these actions are, and how it will affect the final construction, is part of the welder's initial consideration in designing the weld and choosing the correct welding process for a particular job.

These chemical and metallurgical actions are complex in themselves, but with the added influences such as personal welding techniques, potential contamination, influx of moisture and air movement around the weld, to name but a few, it's not surprising that welds occasionally fail.

Most weld failures can be attributed to:

▶ Improper design of weld joint
▶ Poor selection of base materials and filler materials
▶ Inappropriate welding processes

- Residual stresses
- Ineffective or non-existent inspection procedures
- Welded components operating outside their safe parameters

12.1.2 Types of weld failures – cracks

Cracks are a significant element in weld failure. This is because the crack has the potential to grow into a total failure once the weld is under load conditions.

There are a number of reasons why a crack will present itself within a weld. These include poor welding technique, faulty welding equipment and poor joint design. Furthermore, there can be porosity within the weld, or even an amount of slag which can encourage cracks to develop.

Finally, the welding operation itself introduces stress in the vicinity of the weld, with intense localized heating which causes expansion and contraction that can itself cause cracking.

Hot cracks and cold cracks

Cracks in welds can be classified into two types, hot cracks and cold cracks. This refers to the condition of the weld when the crack first presents itself.

Hot cracks, as the name suggests, develop when the weld is still at a high temperature. Initially, they propagate between the grains of the weld material whilst it is still molten and develop into a full crack during the solidification process.

Cold cracks develop at low temperatures, generally because of stresses which develop as the weld solidifies. The influx of hydrogen during the weld process can also have an embrittling effect (loss of ductility) which can encourage cold cracking.

Cold cracks are also known as delayed cracks.

12.1.3 Porosity

Weld porosity is the presence of air bubbles within a weld. If seen through a microscope, the surface of a weld which suffers from porosity would be full of holes. However, those holes are not only on the surface, but they are likely to continue throughout part or even the whole of the weld.

Porosity is a weakening factor in a weld's integrity. If a weld has been designed at the outset to work within certain parameters, then it is indeed possible that an unexpected weld failure will occur under normal working conditions if porosity within a weld is not recognized before the component goes into service.

Porosity happens when the molten weld pool absorbs certain gases (nitrogen, oxygen or hydrogen) which become trapped and result in tiny bubbles once the weld solidifies. The main reason for this absorption is poor gas shielding during the welding process. Gas

shields can also be contaminated by draughts of air around the welding process which can break through the shield and introduce any of the alien gases to the weld process, resulting in porosity.

Contamination from hydrogen – which stems from moisture – can be attributed to moisture being present on the electrodes, the fluxes or the components being welded.

12.1.4 Weld profile

Weld profile refers to the size of the weld.

Weld profiles can have a significant influence on the integrity of a weld. If a weld is too small in section then its load carrying capability will be limited and may well not meet the parameters expected of the finished component.

However, if a weld is too big this can also create problems which may lead to a weld failure. Too much root penetration, which is associated with welds regarded as too big, can create defects and cracks. Excessive weld profiles can also distort components to the extent that if those components are subject to a compressive load when put into service, they could buckle under the strain.

12.1.5 Weld testing techniques

Weld testing techniques – Non-Destructive Testing (NDT)

Given that there are so many factors which can affect the integrity of a weld, it makes sense to use a system of inspection and/or testing at appropriate points in time to ensure any defects are detected before they manifest themselves in a failure. The use of Non-Destructive Testing (NDT) has proved to be a valued procedure in reducing the number of weld failures occurring on components which are in service.

As we've just mentioned, NDT should be conducted at appropriate points in time, those being:

- Once the component has been fully fabricated to ensure the welds do not have unacceptable defects.
- On a regular basis during the service life of the welded component to ensure that no unacceptable defects have developed since fabrication or since the last non-destructive testing was conducted.

There are a number of non-destructive testing methods that can be used. These include:

- Magnetic Particle Inspection (MPI)
- Dye Penetrant Inspection (DPI)
- Ultrasonic flaw detection
- Radiography testing

We will now look at each of these methods in more detail.

Weld testing techniques – Magnetic Particle Inspection (MPI)

Magnetic Particle Inspection (MPI) is a non-destructive testing process which can detect surface or near sub-surface defects in ferrous metals. It does this by first of all putting a magnetic field into the material about to be tested. Once this has been done, ferrous iron particles, which have previously been coated in a fluorescent dye, are introduced to the magnetized area.

Where a surface or sub-surface abnormality (defect) exists, the magnetic flux will leak out, allowing the iron particles to be attracted to that location. This is known as an indication. At this point in the testing process, an ultraviolet light is shone onto the area being tested and this illuminates the fluorescent-coated particles, which indicates the magnetic flux leakage caused by the presence of a defect. The defect can then be analysed as to its severity and what action, if any, is required to rectify the situation and avoid a failure.

Weld testing techniques – Dye Penetrant Inspection (DPI)

Dye Penetrant Inspection (DPI), also known as Liquid Penetration Inspection (LPI), or Penetrant Testing (PT), is a low-cost, non-destructive testing process which can detect surface-breaking defects in non-porous materials. The principle behind the process is that dye will penetrate into any surface-breaking defect and highlight it. Because of its relative low cost, it is an effective way of detecting defects in a wide range of components which are in service, as well as newly constructed fabrications.

Weld testing techniques – ultrasonic flaw detection

Ultrasonic flaw detection is a non-destructive testing process which uses energy waves (high frequency vibrations – ultrasound) to read the exact structure of a component and thus detect any flaws or defects that may be present.

Its main advantages are:

▶ It can detect defects which are deep inside any component.
▶ It can detect very small defects.
▶ It offers a greater accuracy of analysis than other non-destructive testing systems.
▶ In certain circumstances, the size, shape, orientation and location of the defect can be established.
▶ The procedure is non-hazardous.

However, there are some disadvantages, which include the fact that the operator has to be highly trained and experienced not only to operate the equipment effectively, but also to fully understand the readings from the test procedure, which are displayed in a spiked graph form. Also, components which are irregular in shape, are very thin or have a rough surface can be difficult to analyse.

Weld testing techniques – radiography testing

Radiographic Testing (RT) is a non-destructive testing process which uses short-wavelength electromagnetic radiation to penetrate the material being inspected and highlight any defects.

This electromagnetic radiation is emitted from one side of the component and detected and measured on the opposite side of the component. This allows an analysis of the composition of the material to be made. This is usually in the form of a film or negative (X-ray).

However, developments in this form of Non-Destructive Testing (NDT) have evolved into real time radiography. This is where images are reproduced on a monitor screen and can be displayed live as a positive image rather than as a negative X-ray image.

Safety controls and measures when conducting radiography testing processes

Radiography testing involves using intense radiation sources which can expose people who conduct the testing, as well as those in the vicinity, to significant amounts of radiation. Consequently, radiography testing should be conducted under a permit-to-work system which sets out the control measures to be put in place and the checks that must be carried out.

Prior to the work commencing, a risk assessment should be conducted. This should include considering the possibility of a radiation accident occurring and how it can be prevented, or how the consequences can be mitigated by establishing emergency procedures and designating an appropriate person to administer them if needed.

If the testing is to be conducted on readily movable components, this should be done in an adequately shielded enclosure or cabinet. Conducting radiography in this way should always be the first choice wherever possible.

Where it is not possible to place the components to be tested within a shielded enclosure or cabinet, such as onsite fabrications, personal exposure and risks should be controlled by using local shielding, safe systems of working, and applying Radiation Controlled Area designation.

There may also be local regulations regarding giving notice to enforcing authorities that radiography testing is being conducted.

Those conducting the testing should use dose rate monitoring instruments so they can confirm that sealed sources of radiation have fully retracted into their containers once the testing has concluded, or that X-ray

12

sets have stopped emitting radiation. In addition, the use of personal electronic alarming dosemeters can give immediate warning of high dose rates, but these should be used only as an addition to dose rate monitoring and not as an alternative.

REVISION QUESTIONS FOR ELEMENT 3 CONTINUED

4 **Identify THREE** types of non-destructive testing of welds.

You can compare your answers with our guidance at the end of the book under the section entitled 'Revision and assessment guide'.

CHAPTER 13

Element 3

Sub-element 3.3:
Safety critical equipment controls

This chapter covers the following learning outcome

Outline the controls available to maintain safety critical equipment.

13.1 Emergency Shutdown (ESD) equipment and systems

13.1.1 Emergency Shutdown (ESD) systems

The petrochemical industry, both onshore and offshore, processes large quantities of hazardous material within a contained environment. Consequently, it needs to have in place systems which will either prevent loss of containment from happening or mitigate the consequences of such an event if it does happen. These systems are known as Emergency Shutdown (ESD) systems and take the form of various components, each of which is designed to deal with a particular scenario and bring it under control in a safe and effective manner.

Emergency shutdown systems – functionality

All emergency shutdown systems need to have a high Safety Integrity Level (SIL). Safety integrity levels are a measure of the dependability of a system or component. We shall be looking at safety integrity levels in more detail, later in this chapter.

A typical emergency shutdown system is made up of:

▶ Various sensors to detect any fire or escape of gas or vapour
▶ Valves and trip relays to isolate sections of the process
▶ A system logic for processing any incoming signals
▶ An alarm system to warn operators and control room staff of a potential adverse occurrence

The system processes the incoming signals and activates output commands in accordance with the cause and effect chart which will have previously been defined for that particular refinery or installation.

Emergency shutdown systems – typical actions

▶ Shutdown of part systems and equipment
▶ Isolate hydrocarbon inventories
▶ Isolate electrical equipment
▶ Stop hydrocarbon flow
▶ Depressurize/blow down
▶ Activate fire-fighting controls (water deluge, inert gas, foam system, water mist)
▶ Activate emergency ventilation control
▶ Close watertight doors and fire doors

As we have just mentioned, the emergency shutdown system operates using a cause and effect strategy. This strategy allows the system to apply an appropriate response to a given situation.

For example, the activation of one low-level gas alarm in an area processing hydrocarbons would not warrant a full shutdown of the process system. An alarm warning in the control room of the possible presence of a low gas level would be sufficient to allow the situation to be investigated.

However, if the situation was that two high-level gas alarms were triggered (two out of three alarms triggered – known as voting), this would warrant additional responses from the emergency shutdown system in order to bring the process and/or the area to a safe condition.

Emergency shutdown systems – the voting system

To ensure any emergency shutdown system has the highest integrity level, the system is split and usually uses a triplicated microprocessor logic system. This triplicated system ensures that if one microprocessor fails, it does not compromise the implementation of the emergency shutdown system if required.

Each system has three microprocessor channels which provide independent logic processing of the input signals. These are then passed to the output circuits. These outputs are displayed as lights on the fire and gas detection panel within the control room.

The system is based on an analysis of the information by voting on the inputs of signals received from fire and gas detectors positioned throughout the facility.

Actions taken will depend on a voting system of input signals from the detectors. One vote out of three will raise an alarm which will be investigated. Two votes out of three will activate the Emergency Shutdown (ESD) system.

13.1.2 Components of an emergency shutdown system

An emergency shutdown system is made up of many Safety Critical Elements (SCEs). These might include:

▶ Remotely Operated Shut-Off Valves (ROSOVs) – Emergency Shutdown Valves (ESDVs)
▶ High Integrity Pressure Protection System (HIPPS) valves
▶ Deluge systems
▶ Fire and gas detection systems
▶ Vent and blow down system

We will now take a more detailed look at each of these elements.

Components of an emergency shutdown system – Remotely Operated Shut-Off Valves (ROSOVs) and Emergency Shutdown Valves (ESDVs)

Remotely Operated Shut-Off Valves (ROSOVs) are also known as Emergency Shutdown Valves (ESDVs). If required, these valves can be triggered and operated by the emergency shutdown system. The principle of these valves is to isolate sections of the process or inventory quickly in order to reduce the amount of hydrocarbon available to feed a fire or leak.

This type of valve usually fails SHUT using a hydraulic/ spring actuator as an operator. That is to say, it requires power to remain open, but its default position without power is in the SHUT position. This means that if, for some reason, power was not available during an emergency situation, the valve would automatically return to a closed position.

Components of an emergency shutdown system – High Integrity Pressure Protection System (HIPPS) valves

A high integrity pressure protection system is a series of gas-tight shutoff valves which are designed to protect a process plant from over-pressurization. When triggered,

Figure 13.1 Emergency shutdown valve
Source: iStock.

these valves isolate the process plant from the sources of high pressure before the safety parameters of the plant are exceeded, thus preventing loss of containment through a potential rupture. The valves are designed to respond rapidly to a high pressure input signal and will usually complete their activation cycle within 2 seconds of receiving a command signal.

Components of an emergency shutdown system – deluge systems

Water deluge systems are a means of fighting a fire with large amounts of water. They are generally positioned in areas where hydrocarbons are processed or stored, as well as areas where there is the potential for an uncontrolled release of gas which could result in a fire or an explosion.

They work by means of a series of nozzles connected to a piping network. The nozzles are kept permanently in the OPEN position. When an emergency situation arises, the deluge system is activated by the emergency shutdown system, or by one of the other means of activation listed below. This switches on the dedicated fire water pumps, resulting in water being emitted from the nozzles and deluging the area. Figure 13.2 shows a fire deluge system in action at a road tanker loading area.

Figure 13.2 Fire deluge system
Source: Frontline Fire International.

A deluge system can be activated by the following means:

▶ Emergency Shutdown (ESD) system
▶ Electrical (fire detectors)
▶ Pneumatic (glass bulb detectors)
▶ Hydraulic (glass bulb detectors)
▶ Manually

Components of an emergency shutdown system – fire and gas detection systems

The fire and gas detection system is the sensitive part of the emergency shutdown system. It senses if and where any potential adverse activity has occurred and relays this information to the control room.

It is made up of a number of detectors, each of which is specifically selected and strategically positioned so as to raise awareness of an adverse activity at the earliest opportunity so that appropriate action can be taken.

The fire and gas detection system is made up of several types of detectors linked to the emergency shutdown systems. These detectors might include:

▶ Gas detection (toxic and flammable)
▶ Vapour detection
▶ Smoke detection
▶ Heat detection
▶ Flame detection

The type of detector used must be fit for purpose and suitable for the type of gas or fire likely in the area it is covering. It also needs to cover the area or location where any gas leak or outbreak of fire is likely to occur.

As we've mentioned, detectors are generally connected to the remote control room. However, there are occasions when alarms may activate emergency evacuation alarms as well, or simply activate a warning siren or lights.

Components of an emergency shutdown system – vent and blow down system

An essential part of any process system is the vent and blow down system. This is a system that is used to relieve pressure within a process and deal with it in a safe and controlled manner. The pressure relief may involve liquid or gas, depending on the process and/or the situation.

Blow down is normally instigated manually, for example when a system is being prepared for maintenance work and pressure has to be relieved to allow work to commence.

Under normal conditions, hydrocarbons which have been taken out of the process to relieve pressure go to a closed drain header and via the Gas Zone drum (GZ) in order to separate the hydrocarbon liquids from the gases and route the gases to the flare system.

However, blow down also has to deal with emergency situations, such as when pressure needs to be relieved

13

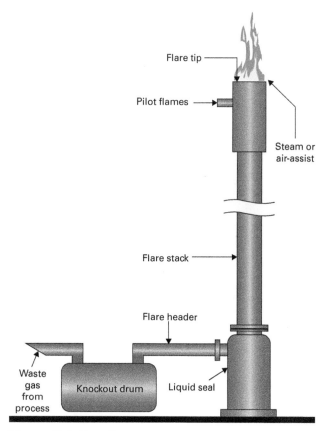

Figure 13.3 Vent flare
Source: Wise Global Training.

because a system has exceeded its maximum safe working pressure. In these cases, the process system will relieve pressure automatically via pressure relief valves and then the blow down system will deal with the exodus of liquid and/or gas.

The type of emergency blow down facility will depend upon the product being processed and the process itself. If the process is of the type that any pressure relief will only involve gas, this might be sent directly to a flare. However, if liquids are involved the system may have to be designed to send these to a dedicated storage tank to be dealt with at a later point.

13.1.3 Level of shutdown

As we mentioned earlier, the level of shutdown needed in response to inputs from an adverse situation needs to reflect the severity of the situation. However, the requirements of each plant, installation or platform will vary accordingly and the shutdown level hierarchy will reflect this.

To demonstrate this we have set out two shutdown level hierarchy charts, one for an offshore platform and the other for an onshore process plant (Tables 13.1 and 13.2).

Table 13.1 Offshore platform shutdown level response hierarchy

Shutdown level	Cause	Effect
Level 1	Failure of any non-critical equipment	Affected equipment shutdown and standby equipment started. Minimal effect on production.
Level 2	Failure of any critical equipment	Production terminated whilst equipment is replaced. Sections of system are isolated to allow equipment to be replaced. Blow down and venting may be required.
Level 3	Fire or gas alarm activated	Production may be affected depending upon the location and extent of hazard. Equipment may be shut down or isolated depending upon which zone is affected and the sensitivity level in that zone.
Level 4	Manual initiation of emergency shutdown	Total production shutdown as well as non-essential utilities being shut down. Product flow from wellhead isolated. Prepare to Abandon Platform Alarm (PAPA) initiated to muster personnel to lifeboat stations.
Level 5	Manual initiation of abandonment	Platform abandon status adopted, inventory vented, Subsea Isolation Valves (SSIVs) shut and all essential utilities shut down.

Source: Adapted from www.hse.gov.uk/comah/sragtech/techmeascontsyst.htm; www.hse.gov.uk/pubns/chis2.pdf

Table 13.2 Onshore process plant shutdown level response hierarchy

Shutdown level	Type of shutdown	Response
Level 1	Unit shutdown	This level involves the shutdown of single process or function. It can be activated automatically by sensors or by manual means. Its aim is to prevent the equipment from operating outside its safe operating envelope.
Level 2	Process shutdown	This level involves shutting down all process systems. It is activated automatically by sensors. Its aim is to isolate all process equipment and limit the potential of an escalating emergency situation developing.
Level 3	Emergency shutdown	This level involves shutting down and isolating specific process equipment through the activation of emergency shutdown valves. It can be activated automatically, or by manual means. Its aim is to limit the consequences during an emergency situation.
Level 4	Emergency depressurization shutdown	This level involves shutting down, isolating and depressurizing equipment and processes by way of opening valves to allow blow down and venting to occur. It is activated manually. Its aim is to reduce the potential for overpressure and to release hazardous vapours and gases from the process.

Source: Adapted from www.hse.gov.uk/comah/sragtech/techmeascontsyst.htm; www.hse.gov.uk/pubns/chis2.pdf

13.2 Safety Integrity Levels (SIL) for instrumentation

13.2.1 Safety Instrumented Systems (SIS)

One of the most critical parts of any emergency shutdown system is the Safety Instrumented System (SIS). A safety instrumented system can be defined as 'a system made up of sensors, logic solvers and actuators which has the ability to take a process to a state of safety when predetermined parameters are exceeded or safe operating conditions are breached'.

An example of this might be the sensor (e.g. gas or flame sensor), the controller (the computer logic solver) and the actuator (the electrical device which activates the emergency shutdown device). This is also known as a safety loop.

All of these individual components, and the system as a whole, need to have a level of integrity or dependency which is in line with the consequences of failure. The greater the consequence, the higher level of integrity or dependency needed.

For example, where a loss of containment due to high pressure could lead to an explosion, the functioning of the High Integrity Pressure Protection System (HIPPS) is critical in maintaining a safe environment. The safety instrumented system (safety loop) associated with the HIPPS must have a high safety integrity level to ensure it will work when required.

13.2.2 Safety Integrity Level (SIL)

Now we have covered what a basic safety instrumented system involves, we need to understand the levels of dependability or integrity that may be required. This is known as Safety Integrity Levels (SIL).

There are four discrete integrity levels: SIL1 to SIL4.

The higher the safety integrity level, the higher the associated safety level requirement is, and this needs to be coupled with a lowering of the probability that the safety instrumented system will fail.

The probability of failure is defined numerically as follows:

▷ SIL 1 – this level of integrity is where the acceptable probability of failure is between 1 in 10 occasions and 1 in 100 occasions. It is required where the potential for relatively minor incidents is involved with limited consequential outcomes.

▷ SIL 2 – this level of integrity is where the acceptable probability of failure is between 1 in 100 occasions and 1 in 1,000 occasions. It is required where the potential for more serious, but limited incidents is involved and where the consequences may result in serious injury or death to one or more persons.

▷ SIL 3 – this level of integrity is where the acceptable probability of failure is between 1 in 1,000 occasions and 1 in 10,000 occasions. It is required where the potential for serious incidents is involved and where the consequences may involve a number of fatalities and/or serious injuries.

▷ SIL 4 – this level of integrity is where the acceptable probability of failure is between 1 in 10,000 occasions and 1 in 100,000 occasions. It is required where the potential for a catastrophic incident exists.

The required safety integrity level for each safety instrumented system is established by conducting a hazard and risk assessment (hazard and operability study or hazard identification study). This will determine the consequences of the safety instrumented system not functioning as required, which in turn will indicate the appropriate Safety Integrity Level (SIL) for that safety instrumented system.

Once the safety integrity level has been established, it is necessary for those who are designing the safety instrumented system to demonstrate that the required level of integrity has been achieved. The design process itself may well go through a hazard and operability study or hazard identification study assessment in order to analyse the different potential scenarios of failure of the safety instrumented system. This will allow contingencies to be built into the system such as duplication of critical devices such as microprocessors.

13.3 Procedures for bypassing Emergency Shutdown (ESD) systems

13.3.1 Bypassing ESD

All emergency shutdown systems and fire and gas systems need to be tested, inspected and/or maintained on a regular basis to ensure they are functioning as required.

These testing and/or maintenance procedures involve the temporary bypassing of safety system interlocks. These are the devices within the logic system which activate the alarm as well as sending a signal to the actuator on the emergency shutdown component itself.

The procedure of bypassing a safety system interlock is known as applying an 'inhibit' or an 'override'. However, by inhibiting or overriding a safety system in order to test it, this in effect compromises the integrity of any emergency shutdown system during the period the inhibit or override is in place. Consequently, everything possible must be done to ensure the duration of time the safety system interlock is out of action is kept as short as possible. Furthermore,

13

inhibits should only be applied as part of an established procedure, which should include conducting a risk assessment prior to applying the inhibits in order to evaluate the potential risks during the period the emergency shutdown system is not protected by the safety system interlock.

Inhibits, generally speaking, should only be applied to one protective function at a time and not to groups of protective functions. The total number of inhibits in place at any one time should not be excessive and should always be manageable. Inhibits should be removed as part of the process of finishing the relevant testing or maintenance activity.

The control room operator should also be aware at all times of the application and location of all inhibits. Furthermore, where inhibits are applied through a visual display unit (VDU), there should be a protective system in place which requires a password or key to be entered before an inhibit can be applied. This will prevent any accidental or unauthorized application.

Finally, where an installation is controlled from two different locations, such as a Normally Unmanned Installation (NUI)/remote station onshore, the inhibits which are in place should be visible at both locations. However, where the control of the inhibits moves from one control location to the other, a proper handover protocol should be applied.

Inhibits should only be applicable to protective functions which are graded to Safety Integrity Levels 1 (SIL1) and Safety Integrity Levels 2 (SIL2). Protective functions which are graded Safety Integrity Levels 3 (SIL3) and Safety Integrity Levels 4 (SIL4) should not normally be capable of being inhibited, as the loss of their safety function is generally regarded as unacceptable.

Bypassing ESD – logging actions and displaying warnings

Whenever an emergency shutdown system or fire and gas system is bypassed, the time the inhibit or override was initiated must be recorded. This is usually covered on a Protective Systems Isolation Certificate (PSIC) or a permit-to-work system, and the inhibit or override is logged and recorded in the central control room. The particular keys used to inhibit or override a system should also be identified and recorded on the inhibit log until the work is completed and the Protective Systems Isolation Certificate (PSIC) or a permit-to-work system is cancelled and the inhibits or overrides are removed.

When inputs or outputs of an emergency shutdown system are inhibited, they should be allocated a low priority alarm and be visible on the display screen of the control room operator. There should also be a visible indication on the appropriate emergency shutdown or fire and gas panel to show that the system has been inhibited or overridden.

The restoration of the safety system interlock to its normal operating condition will clear the alarm.

All inhibits or overrides should be logged and a record kept. An inhibit log should record the following information:

- The safety function being inhibited
- The time and date the inhibit was applied
- A cross reference to the relevant permit-to-work or protective systems isolation certificate where applicable
- The time and date of each reassessment
- The time and date the inhibit was removed

Other control measures to be applied to inhibits and overrides include:

- They should only be applied with the authority of a senior person or manager.
- They should be time bound at the outset. Equipment must be shut down if the emergency shutdown system is not reinstated within the designated time.
- Their implementation should be communicated to all operation personnel as well as adjacent plants.

13.4 Blow down facilities, flare types

The term 'blow down' refers to the action of venting gas or relieving pressure from a process or production system. For example, if a system reaches its maximum working pressure and an automatic pressure relief valve vents gas from the system, then the blow down facility will take the gas away to a safe exit point where it will be allowed to escape into the atmosphere, or to a flare where it will be burnt off. This also happens if gas is released during an emergency shutdown event.

Blow down can also be instigated manually, for example when a system is being prepared for maintenance work and pressure has to be relieved to allow work to commence.

When controlling overpressure, it is not uncommon to have both a high pressure and a low pressure vent system. With regards to high pressure vessels, any blow down or depressurization can be a highly hazardous operation. This is because any rapid reduction in pressure can lead to a reduction in a vessel's wall temperature with the possibility of reaching a temperature below its ductile-transmission level and thus the potential of a brittle fracture occurring in the vessel wall.

Another consideration of a rapid reduction in pressure is the formation of hydrates if free water is present in the vessel or associated pipework.

Other considerations in the operation of blow down or venting systems include:

- The reduction in temperature of the system and pipework as it relieves the pressure/gas

- The noise levels during venting
- The maximum blow down rates and times
- The liquid which is knocked out of the gas as it vents
- The appropriate blow down and venting system to be used which meets the type of pressure on the system being relieved (high pressure and low pressure systems being split and independent)
- The system relieving capacity

Most vent stacks consist of a single pipe which rises above the installation to an elevation which is regarded as safe for the gases to be vented to atmosphere. The system may also have within it a 'knock out' facility which will remove any liquids which are suspended in the gas, before it is vented away.

Finally, in the case of a lightning strike hitting the vent system and potentially igniting the gas within it, there is usually a form of snuffing system built into it. This may use CO_2 or steam as a means of extinguishing any flames that are caused by the lightning strike.

Figure 13.4 Two flare stacks
Source: iStock.

Flare stacks are used to burn off flammable gas when this is released during planned and unplanned events on a process or production system.

They are also used in refineries to burn off purged and waste products, as well as on drilling rigs to burn off unrecoverable gases which can accompany oil extracted from the wells.

13.4.1 Types of flares

Flares work on the basis of introducing air at the point where fuel gas is combusted so that efficient and complete combustion is achieved and smoke is kept to a minimum. Exactly how this is achieved will depend on the characteristics of the fuel gas, including its composition and velocity.

Let's look at some of the different types of flares used.

Steam assisted flares

These types of flares are the most prevalent in the oil and gas industry. They employ a system which injects steam into the flare at the flare tip. This introduces air into the flame, which makes the flame more efficient whilst at the same time reducing the tendency for the flame to emit smoke. This is an important consideration where installations are close to populated areas. However, injecting steam does increase the noise level emitted by the flare, which can be a problem.

Air assisted flares

Where waste gas pressure is low and steam is not available, then air assisted flares can be used. This type of flare uses a fan to direct air flow to the flare tip. This air flow assists in two ways. First it adds momentum to emission of the fuel gas, and second it provides a contribution towards the air needed for efficient combustion.

Unassisted flares

Unassisted flares use only the pressure of the emission of the fuel gas to provide momentum for the flame. The pressure of the fuel gas helps mix the gas with air for combustion.

Multi-point pressure assisted flares

Multi-point pressure assisted flares use the principle of high exit velocity (at or near sonic velocity) of the fuel gas from each burner. This induces the mixing of fuel gas with air for efficient combustion. Generally each burner is approximately 2.5 metres tall and they are arranged in rows, the formation of which are known as a Multi-Point Ground Flares (MPGFs); the area they cover can be as big as a football field. Some of these multi-point flare systems are capable of handling over 4,400 metric tonnes of gas per hour.

In order to cope with the variation of gas volume and still maintain the high exit velocity required, each row of burners is fed by a staging valve. When more fuel gas flows, more valves are opened, and when less fuel gas flows, some of the valves are closed. This ensures that sufficient back pressure is maintained to facilitate the high exit velocity.

13

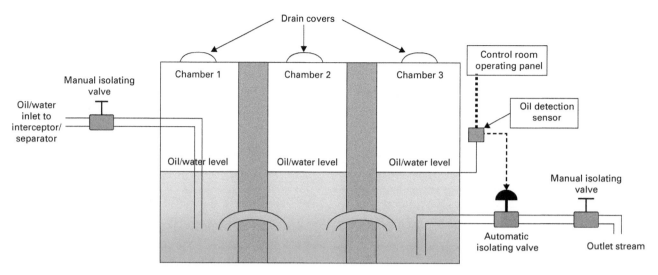

Figure 13.5 Interceptor separator
Source: Wise Global Training.

13.5 Closed and open drain headers, sewers, interceptors

13.5.1 Drainage

With any process or production facility there will always be a residual amount of liquid that finds its way to ground, or there will be a requirement for vessels to be occasionally drained of liquids of some sort so they can be maintained. These requirements have to be accommodated in some way, and the normal method is to provide a drainage system in which the fluids can be collected and dealt with appropriately.

Drainage systems are a means of safely collecting any residual process fluids, hydrocarbon liquid and/ or chemical spills, deluge water and rainwater, and transporting it to an appropriate location where it can be dealt with or disposed of appropriately. There are two types of drainage systems: open drains and closed drains. We shall now look at both types in more detail.

Drainage – open drain systems

A drainage system that collects fluids that spill onto the ground is called an 'atmospheric drain', 'gravity drain' or 'open drain'. This system will consist of collecting funnels or trays known as 'tundishes' which are strategically positioned so as to channel any liquid they collect through a series of drain pipes to an open drain header before being routed to a slop tank ready to be dealt with appropriately. Water seals should be fitted on drainage pipes where they transport liquids from one location to another so that they do not act as a conduit for any gas to also travel through the system.

There will be separate drainage systems to deal with 'safe areas' and 'hazardous areas'.

Drainage – closed drain systems

A drainage system that is required to be connected to a pressure vessel in order to drain off fluids is called a 'pressure drain system' or 'closed drain system'.

Whenever a vessel is drained of liquid under pressure into a pressure or closed drain system, it must be assumed that the liquid contains a certain amount of dissolved gases. Furthermore, the flow of liquid from the pressurized vessel will be followed by a certain amount of gas (known as gas blow by). This gas, in both its forms, can represent a hazard if it is not dealt with appropriately, e.g. using a blow down facility so that gas is vented away to a flare.

Also, because of the risk this gas represents, a closed drain should never be connected to an open drain system.

Drainage – interceptors

The draining of contaminated water to sewers and/or groundwater must be avoided. Interceptors are a means of collecting contaminated water before it is discharged to a foul drain or surface drain.

Typically, interceptors have three separate chambers with the divisions between chambers extending down to the bottom, and low level pipes connecting the chambers. This is so that when the contaminated water enters the first chamber it can separate (oil will naturally float on top of water) and be extracted. The water is then directed to the second and third chambers via the low level pipe, where any residual oil is also allowed to separate and be extracted. Finally, the water from chamber three is channelled into either a foul or surface drain, whichever is appropriate.

Interceptor systems need to be emptied of residual oil and other contaminates on a regular basis. They are normally fitted with robust visual and audible alarm systems to indicate when these operations need to be conducted. They can also be fitted with automatic closure devices fitted to the retention separators to prevent flow passing through the separators when the quantity of oil in the separator exceeds the oil storage volume. However, there should be a high level alarm system in place to warn the operator when the automatic closure device has been activated so that immediate action can be taken.

REVISION QUESTIONS FOR ELEMENT 3 CONTINUED

5 **Describe** what an emergency shutdown system is and its function.
6 **Identify THREE** typical actions an emergency shutdown system might perform if activated.
7 **Explain** why and when an emergency shutdown system might be bypassed.
8 **Describe** the purpose and function of an interceptor.

You can compare your answers with our guidance at the end of the book under the section entitled 'Revision and assessment guide'.

13

CHAPTER

14

Element 3

Sub-element 3.4:
Safe containment of hydrocarbons

This chapter covers the following learning outcome

Outline the hazards, risks and controls available for safe containment of hydrocarbons offshore and onshore.

14.1 Hazards and risks including overfilling, effects of vacuum, overloading of foundations and failure modes for tank shells and associated pipework

14.1.1 Storage tanks

Hydrocarbons are usually stored in tanks like those shown in Figure 14.1 below. Depleted wells and salt caverns are also used as storage facilities.

However, for the purposes of this module we will be concentrating solely on storage in tanks.

The type and quantity of the product being stored will dictate the type of tank used, as well as the measures used to control the risk of it failing. For instance, gases such as propane may require refrigerated or pressurized storage, whereas oils can be stored in standard storage tanks like those shown below. However, the quantity of oil to be stored, together with the oil's particular volatility, are factors which will also govern the type of tank to be used.

Figure 14.1 Storage tanks
Source: iStock.

Storage tanks – capacity

Tanks may be quite small (e.g. 1,000 m³) or very large (e.g. 50,000 m³). Consequently, the foundation on which the tank is built should be substantial enough to take the weight of the tank when full. However, this is not a straightforward calculation as the density, or specific gravity, of oil varies from approximately 0.8 tonnes per m³ to 1.3 tonnes per m³ depending on its type and grade.

Furthermore, it is not just the capacity of the foundations that determines the amount of product that can be stored. The particular design and construction of the tank shell, along with its support structure, will also be a consideration.

Each tank will have a prescribed maximum volume although, as we've mentioned, the density of oil varies greatly depending on its type and grade and this prescribed volume may have to be reduced accordingly for that reason.

During a storage tank's life, the quantity of product in the tank will need to be carefully monitored and this can be done manually by sight gauge, or remotely with gauges that feed back to a central control room. As a secondary measure, these signals are usually linked to an alarm system which will warn of any exceptionally high or low levels.

The risk of overfill is greatest when product is being pumped into the tank from another source, such as a road or sea tanker transporter. The overfill alarm system may well have an auto-shutdown facility, but this is not always the case and dependency on such devices is not recommended as they can fail. Consequently, good communications and standardized procedures between the party sending the product and the recipient is essential.

Storage tanks – integrity management

Storage tank integrity is an important aspect which needs to be well managed and controlled. This is because these tanks contain large amounts of hazardous liquid which, if it leaked, could result in a catastrophic event.

Tanks are often thought of as simple structures that require little attention, and for this reason arrangements generally are not as robust or comprehensive as for other process plant and equipment. Nevertheless, storage tanks can deteriorate in complex and varied ways.

Another consideration is the compatibility of the product being stored with the material the storage tank is made from. Not all products are compatible with all materials and knowledge of what product can be stored within tanks of a specific material is essential. Also, as we've mentioned, the specific gravity of oil differs from type to type and this will affect the safe operating level of the storage tank.

The following management practices, which are regarded as good practice with regard to managing storage tank integrity, are taken from the HSE document 'Integrity of Atmospheric Storage Tanks'.

- Tanks containing hazardous substances should be identified and entered in the plant register.
- Operators should maintain tank data files.
- Compatibility assessments should be undertaken (e.g. tanks on multi-product service).
- Tanks should be subject to formal periodic maintenance and inspection.
- Inspection and maintenance of tanks should only be carried out by experienced and qualified competent persons.
- Schemes of inspection should be established and agreed between operators and competent persons.
- Appropriate inspection techniques should be utilized, depending on deterioration mechanisms.
- Inspection reports and checklists should be of high quality.
- Where necessary, recommendations from inspection reports should be actioned promptly.
- Assessment of fitness for service should be carried out following tank inspection and significant changes to process or operating conditions.
- Operators and competent persons should have knowledge of, and adopt the recommendations given in, the relevant guides, codes and standards.
- Tank examination schemes should include both internal and external inspections.

14.1.2 Evolving damage mechanisms

There are many types of evolving damage mechanisms which can affect storage tanks, many of which can work simultaneously. They include the following:

- Corrosion
- Erosion
- Creep associated with non-metallic thermoplastic tanks
- Fatigue
- Chemical attack
- Brittle fracture

We will now take a more detailed look at each of these evolving damage mechanisms.

Evolving damage mechanisms – corrosion

Corrosion is the gradual deterioration of a substance by way of a chemical reaction with its immediate environment. Corrosion is most commonly associated with rusting, which is when metal reacts with oxygen to form iron oxide. This is a form of electromechanical corrosion.

Most storage tanks are made from carbon steel, which makes corrosion a primary cause of deterioration. Furthermore, corrosion can occur both internally and externally, which makes the monitoring of corrosion a difficult issue.

Tanks known as 'flat bottomed tanks' are particularly prone to corrosion on the underside of the top plate as well as the upper side of the base plate. This tends to be aggravated if the product being stored is experiencing an aqueous phase (aqueous meaning water present within the product). Product temperature is also an important factor in the amount and rate of corrosion.

Other factors which can affect corrosion in the base plate are:

- The materials used in construction
- The condition and the effectiveness of the floor-to-base seal
- The angle of the tank pad away from the base of the tank

The floor of a 'flat bottomed tank' may not be flat. This is because most are cone-shaped so that water and other bottom products can be allowed to gravitate to a point where they can be drawn off. Cone-up bottoms have sumps around the edge of the tank to take bottom products away, whereas cone-down bottoms have a sump in the middle to do the same job.

Bottom products often contain aggressive compounds, and the monitoring of the pH of these products may be regarded as good practice in monitoring and controlling corrosion.

As well as top and bottom plates, corrosion of the lower part of the shell strake, or wall, is a common feature. This may be caused by rainwater being allowed to settle around the tank base. One means of controlling this, apart from ensuring there is good drainage away from the tank, is to paint the bottoms and lower section of the tank wall with a protective coating.

14

We mention in Chapter 11 the issue of annular rim failure, and it's worth repeating part of that issue here because of its relevance.

Annular plates are prone to corrosion attacks both on the outer side where the tank shell sits on the annular rim and on the underside of the annular plate where trapped water may lie undetected. This corrosion, coupled with the prolonged stress, can lead to stress corrosion cracking and failure occurring without warning.

Drains and draw-off sumps used to transport water and bottom products away from storage tanks are also vulnerable to corrosion, and this is compounded by the fact that they are difficult to inspect, and the culverts they lie in are often submerged in rainwater.

Other areas of corrosion are as follows:

▶ Where fittings are welded to the tank shell
▶ Potential water traps, e.g. below wind girders, stairwells and stairway connections
▶ Liquid/vapour interfaces
▶ Around breathers and vents

Another common form of external corrosion is Corrosion Under Insulation (CUI). Here the temperature of the stored product is a critical factor. Some products have to be stored within a range of 50°C–100°C and, as such, the tank needs to be insulated. When this is coupled with one or both of the compounding factors given below, an ideal environment for corrosion is created.

The two aggravating factors referred to above are:

1 If the insulation material is high in chlorides, this will increase corrosion.
2 If the insulation material is not cut away on the bottom edge it will lead to a 'wicking effect'. This is where moisture is drawn up from floor level and retained within the insulation.

Control measures for Corrosion Under Insulation (CUI) are as follows:

▶ Use low chloride or chloride free insulation.
▶ Vulnerable areas, such as the roof-to-shell joint, should be effectively sealed.
▶ Fit supports for the insulation.
▶ Apply small-bore penetrations to the insulation to allow moisture to escape.
▶ Ensure the lower 20–25cm of the tank shell is free of insulation.

When undertaking maintenance or inspections, extreme care must be taken when walking on a tank roof, as its integrity cannot be taken for granted. Also, internal floating roofs tend to be of very lightweight construction and, as such, are not suitable for people to walk on.

Evolving damage mechanisms – erosion

Erosion is the process of material being worn away by the constant movement of product flowing over the surface. Areas such as filling and discharge points which experience large amounts of product flow are the most vulnerable points.

The main control measure for erosion is to increase the thickness of the material where erosion is identified as a threat.

Evolving damage mechanisms – creep associated with thermoplastic tanks

Creep is where the tank material stretches over a period of time when it is under stress from the weight of the product being stored. However, this condition only affects non-metallic thermoplastic tanks.

Increases in temperature tend to compound the problem with High-Density Polyethylene (HDPE) being particularly vulnerable and losing its strength as the temperature increases.

Evolving damage mechanisms – fatigue

Pressure vessel storage tanks are more prone to fatigue defects than tank storage vessels. This is because the membrane stresses in pressure vessels are much greater.

For tank storage vessels which are subject to a cycle of frequent filling and emptying, certain parts of the tank's construction can be vulnerable, particularly welded joints.

Control measures revolve mainly around inspection and monitoring so that any problems can be detected and dealt with before they become an issue.

Dye penetration testing and other non-destructive testing techniques can be used to detect fatigue-induced defects.

In Glass Reinforced Plastic (GRP) tanks, fatigue is a recognized feature and a defined service life for these tanks is standard.

Evolving damage mechanisms – chemical attack

As we mentioned earlier, the compatibility between the product being stored and the material of the storage tank is paramount. If this is not the case, then there is a possibility of deterioration due to chemical incompatibility. Materials stored at elevated temperatures may need additional consideration, as the aggressiveness of some products is greater at these increased temperatures.

The most common materials used in the construction of storage tanks are carbon steel, stainless steel and Glass Reinforced Plastic (GRP). Carbon steel is suitable for a large number of products, although some products with an acid content may react with carbon steel.

Stainless steel is used where purity of the stored product is paramount. However, welded areas in stainless steel tanks can be prone to stress corrosion cracking.

Glass reinforced plastic suffers from ultraviolet light degradation over time and has a reduced ability to perform as product temperature is increased.

Evolving damage mechanisms – brittle fracture

Brittle fracture tends to be most prevalent either during hydro-testing shortly after construction has been completed or during the first filling in cold weather. If these two events have been successfully completed, then experience shows further risk of brittle fracture is minimal.

14.1.3 Mechanical damage

Mechanical damage can be inflicted on storage tanks in many ways, including:

▶ Impact
▶ Settlement which is non-uniform
▶ Over-pressurization
▶ Vacuum
▶ Excessive external loads such as snow, ice and rain, where the load has not been allowed to drain away
▶ Wind loads
▶ Tanks floating off from their foundations

Let's take a look at each of these issues.

Mechanical damage – impact

An impact on a metal tank will have a distorting effect which can compromise tanks with floating roofs. Non-metallic tanks have poor impact resistance and may well fracture if the impact is serious.

Control measures include placing barriers in areas where impacts may occur. Similarly, bunding may act as a protective barrier. Good inspection regimes will also pick up on externally inflicted damage.

Mechanical damage – settlement which is non-uniform

Subsidence can be caused by constant weight on weak or compressible terrain. The weakness or compressibility may not be uniform, which can lead to distortions in the tank structure. Foundations which are designed and built to compensate for local conditions are essential.

Two further causes of foundation movement are frost heave, where the ground is subject to frequent freezing and thawing, and ground movement caused by high tides in areas close to the sea.

A precautionary measure where these factors are recognized as issues is to fit connecting pipes with flexible joints or bellows to allow for movement.

Mechanical damage – over-pressurization

Over-pressurization is where pressure is allowed to build up to unacceptable levels and can lead to

a tank rupture. This is most likely to happen when the tank is being filled and the pressure relief valve fails. Controls include regular inspection and maintenance of these valves and any sensors associated with them.

Mechanical damage – vacuum

Vacuum generally affects fixed roof storage tanks. As the tank is gradually emptied of product, if the relief valve or vent becomes blocked or fails, a vacuum will build up causing the tank to distort. Again, controls include regular inspection and maintenance of these valves and vents together with any sensors associated with them.

Mechanical damage – excessive external loads

The most common external loads which can cause damage to a storage tank are snow and ice. These can cause buckling where the loads are excessive and should be cleared before they cause problems.

Rainwater can also be an issue, especially on floating roofs where drains have become blocked.

Mechanical damage – wind loads

Wind can cause tanks to lift, especially if they are not anchored down or have little product in them.

Mechanical damage – tanks floating off their foundations

Tanks are surrounded by bunds. These serve the purpose of containing product if there is a leak. However, in a situation where rainwater is allowed to build up inside a bunded area with a tank in it which has not got a lot of product within it and which has not been anchored down, it is possible for the tank to float off its foundations.

Controls are to keep bunded areas free of rainwater by pumping it out on a regular basis. There are specialized pumps available now called 'bund water control units' which monitor the rainwater collected within the bund and automatically pump it to an onsite interceptor. These units also have an automatic oil detection function which will shut down the pump if oil is detected and send an alarm signal to the control room.

14.2 Floating roof tanks, landing the roof, sinking the roof and rim seal fires/failures

14

14.2.1 Introduction

Storage tanks play a major part in the oil and gas industry, where large amounts of hydrocarbons have to be stored at various stages of their process to become a useable product, or are awaiting transportation for final distribution.

As we mentioned earlier in this chapter, storage tanks come in many different sizes and designs, and the size and design is dictated by the quantity and type of product they are expected to store.

There are tanks which have a fixed roof and those which have a floating roof. We will look at those tanks with floating roofs first.

Tanks with floating roofs can have an 'internal floating roof' or an 'external floating roof'.

Storage tanks with what are known as 'external floating roofs' have a containment cover which floats inside the shell walls, but is open to the elements.

Storage tanks with an internal floating roof still have a containment cover which floats inside the shell walls, but has a roof covering over the outer shell.

14.2.2 External floating roof tanks

An external floating roof tank is made up of a cylindrical shell with a fixed base. This shell then has a roof within it which floats up and down on top of any liquid stored within the tank. This means that there is no vapour

space – known as ullage – in a floating roof tank. This lack of ullage in a floating roof tank reduces product loss due to evaporation. A rim seal between the roof and the inner tank wall also ensures evaporation is kept to a minimum.

The exception to this is that, if there is a very low level of liquid in the tank, the legs on the underside of the roof will engage with the base and a void will evolve. This is an aspect called 'landing the roof', and we shall be looking at this later in this section.

The main disadvantage of an external floating roof is that snow can accumulate on the roof and can build up to such a level as to cause the roof to sink. The same can happen with rainwater if the drain from the roof becomes blocked. Also, the rainwater on the roof is usually drained through a hose which passes through the stored product to a drain valve in the base of the tank. If this hose develops a leak, it will drain both rainwater and product.

14.2.3 Internal floating roof tanks

An internal floating roof tank uses the same principle as an external floating roof but the tank then has a fixed roof covering the tank, as shown in Figure 14.3 below.

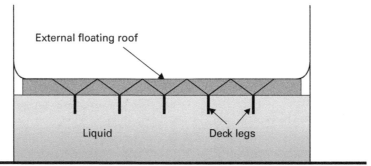

Figure 14.2 Section through external roof storage tank
Source: Wise Global Training.

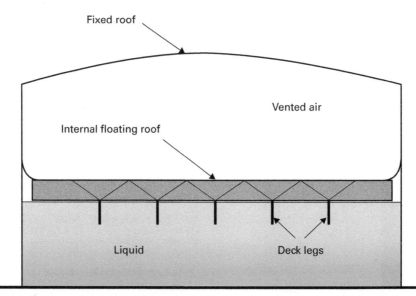

Figure 14.3 Section through internal roof storage tank
Source: Wise Global Training.

Aspects relating to internal floating roofs tanks are as follows:

▶ The internal floating roof tends to be made of a very lightweight construction and, as such, is not suitable for people to walk on.
▶ Internal floating roof tanks are free venting. That is to say, they are not fitted with Pressure and Vacuum Valves (P&Vs).
▶ Internal floating roof tanks tend to be used for storing material with a low flash point, such as gasoline.

14.2.4 Issues with floating roofs

There are issues to be considered with both types of floating roof system. These include:

▶ Landing the roof
▶ Sinking the roof
▶ Rim seal failures and rim seal fires

We will now look at these issues and their ramifications.

Issues with floating roofs – landing the roof

Landing the roof is where the liquid in the tank falls far enough for the legs on the underside of the floating roof to land on the base of the tank. The void between the liquid and the roof will grow, which will allow a build-up of vapour to occur. This has the potential to cause a fire and/or an explosion.

If the roof is landed, it is important that enough time elapses before the tank is refilled to allow the space below the roof to become saturated with vapour so that it exceeds its upper flammable limit. This is because when product is pumped into the tank and the roof begins to lift, there is a chance that a spark may be generated between the tank wall and the roof seal. Having the vapour in the ullage saturated to beyond its upper flammable limit will ensure the vapour will not be ignited by any errant spark. The amount of time required to allow the vapour to become saturated may be as much as 24 hours.

Another issue with landing the roof is that it causes a particular corrosion mechanism called fretting-related corrosion. This is where repeated contact with the base of the tank by the legs on the underside of the floating roof tends to remove protective layers of rust scale from the base that may have formed over time. This has the effect of increasing corrosion rates at these locations.

Issues with floating roofs – sinking the roof

Sinking the roof can be caused when the roof becomes unbalanced for some reason and sinks into the liquid. Reasons for becoming unbalanced include:

▶ Build-up of rain or snow
▶ Use of access ladders on one side of the roof
▶ In certain parts of the world, earthquake activity may have a destabilizing effect on the roof

With internal roofs, if the roof sinks the ullage space will quickly fill with flammable vapours and create a fire/explosion hazard. However, the volatile atmosphere will be contained within the tank beneath the fixed roof.

With external roofs, if the roof sinks, the flammable vapours are free to evolve and drift away and may meet a source of ignition.

Issues with floating roofs – rim seal failures

Floating roofs move up and down within the shell of the tank according to the level of liquid in the tank. The space between the tank wall and the roof is kept airtight by a rim seal which ensures vapours don't escape, and contaminates such as rainwater don't get into the stored liquid.

Most systems have a double seal so that there is a back-up in place should the primary seal fail. Furthermore, there is usually a detection system built onto the roof so that if a leak does occur, it can be detected and dealt with. The detection system may well incorporate an automatic rim seal fire-fighting system as well.

Without these safety features, vapours could escape unnoticed. In such cases there have been occasions where an electrical storm has created a set of circumstances which have created a spark adjacent to the tank. This has subsequently ignited the vapour and caused a fire.

All tanks should be well earthed and lightning conductors erected to prevent lightning striking the tank. They should also be frequently checked for earth continuity.

We shall be looking at rim seal fires and how they can be dealt with in Chapter 15.

14.3 Fixed roof storage tanks, pressure and vacuum hazards

14.3.1 Fixed roof tanks

Fixed roof tanks can either be self-supporting or have internal supports for the roof.

Fixed roof tanks undergo numerous loads and pressures during their working life and understanding what these loads and pressures are, and the circumstances in which they occur, is an important part of controlling the hazards they generate.

The weight of any liquid pumped into a tank will naturally apply outward pressure on the walls of that

14

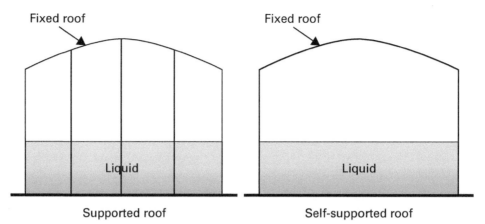

Figure 14.4 Section through fixed roof storage tanks
Source: Wise Global Training.

tank. Also, as the liquid builds up inside the tank, the atmosphere in the void needs to be vented out. If this does not happen the pressure will become unsustainable and the tank may rupture.

The contra view is also true, in that when a tank is emptied the liquid reduces and the void above the liquid needs to have air vented in so that a vacuum is not created. If the air inlet valve is faulty, and a vacuum is created, this may well cause the tank to collapse.

There are other factors which can cause increased or decreased pressure within the tank. These include:

▶ Storing a volatile product will cause gases to evolve and increase pressure.
▶ Warm weather or direct sun on the tank will warm the product up and make it expand, thus increasing pressure.
▶ Cold weather will cool the product down and cause it to contract, thus decreasing pressure.

Normally, tanks have a Pressure and Vacuum (P&V) relief valve fitted on or near the top of the tank. This allows vapour or gas to escape or, in the case of a reduction in pressure within the tank, allows air to enter the tank. This ensures the integrity of the tank is retained.

Consequently, the P&V relief valve is a critical piece of equipment and the integrity of the tank is dependent on it working properly, and as such it should be regularly inspected and maintained.

Another method of controlling changes in pressure is to employ an expansion or pressure exchange system. This is where the expelled vapour or gas is directed to another tank rather than to atmosphere.Where a vacuum has to be dealt with, the system draws in an inert gas from an external source.

This system ensures volatile gases or vapours are not released to atmosphere in an uncontrolled fashion.

14.4 Bunding of storage tanks including volume and area sizing, construction and valving arrangements

14.4.1 Bunding of tanks

Where storage tanks hold a liquid product there is always the possibility that the tank will rupture and spill its contents. Consequently, contingencies should be in place to control such a situation. This is where a containment wall, known as a bund, is built around a tank, or group of tanks, to contain any spillage that may occur until such time that it can be dealt with.

The bund also stops the product spreading too far, seeping into the ground, or seeping into the drainage or water systems. It also reduces (although it does not eliminate) the risk of fire and/or explosion.

We have listed below some of the effects and consequences of a tank failure which is not contained by a bund.

Environmental effects

▶ Contamination of the surrounding ground area
▶ Contamination of water courses should the spill reach public sewers and/or drains
▶ Impacts on wildlife and aquatic life
▶ Evaporating fuel from the spill having an effect on the ozone layer

Consequences and effects

▶ Fire (impingement on other plant and equipment)
▶ Explosion
▶ Uncontrolled release of a substance which may be toxic
▶ Health risks to personnel on site
▶ Contamination with other products
▶ Damage to neighbouring businesses
▶ The risk to personnel onsite

▶ The risk to personnel offsite (local residents)
▶ Flooding into catchment ponds, interceptors, etc. and reaching public sewers, drains and waterways
▶ Business shutdown, loss of revenue, loss of company good name and image

Figure 14.5 Bunded storage tanks
Source: Dreamstime.

Bunds are designed to contain any spillages and stop them escaping. This includes seeping into the ground. Consequently, bunds should be built on an impervious base and the material the bund is constructed from should also be impervious.

A bund needs to contain any amount of spillage from the tank(s) it surrounds. Therefore the capacity of the bund must be equal to the total maximum content of the tank(s) within the bund plus an extra 10 per cent.

The bund itself must be complete, that is to say it must not be breached. Any pipelines must go over the bund rather than through it. Access to the tanks inside the bund must again be over the bund rather than through a gate.

Figure 14.6 Pipes going over bund
Source: Dreamstime.

There should also be some sort of high level alarm warning system on the bund linked to the central control room to warn the control room operators of any leaks.

Should a spillage occur, arrangements must be in place to deal with it. Similarly, arrangements must be in place to remove rainwater which may accumulate within the bund.

Removal of rainwater can be a costly process, as the water may be classed as hazardous waste due to its potential contamination with product. However, there are specialized pumps available now called 'bund water control units' which monitor the rainwater collected within the bund and automatically pump it to an onsite interceptor. These units also have an automatic oil detection function which will shut down the pump if oil is detected and send an alarm signal to the control room.

Tanks can be individually bunded, or bunded as a group. However, where tanks are bunded as a group, it is important to ensure that products within the tank grouping will not conflict with one another if they leak at the same time.

Finally, the bund wall should be substantial enough to maintain its integrity if a fire occurs.

14.5 Filling of tanks, overfilling/alarms/tanker connections

14.5.1 Transfer of product

The transfer of product from one location to another comes with its own inherent hazards. Any transfer operation has the potential of leakage or overfill, and measures must be taken to minimize such risks. We are going to look at transfers involving storage facilities.

Transfers can be from road or rail tankers to a storage facility, or vice versa. They can also be from a marine tanker to a storage facility, or vice versa. We shall concentrate on road/rail transfers here as we shall be looking at the issue of marine transfers in Chapter 20.

14.5.2 Transfer of material between road/rail tankers and tanks

The safe transfer between road/rail tankers and tanks should include:
▶ Securing the road/rail tanker – brakes applied, and engine turned off.
▶ Ensuring that hoses are suitable for the product being discharged and the operating pressures they will be subjected to.
▶ Ensuring the connections of pipes and hoses to be used in the transfer operation are secure.
▶ Positioning of drip trays beneath all connections and ensuring there is close monitoring of connections during transfer operations.

14

- ▶ Making a good bonding connection between the road/rail tanker and the loading/unloading equipment. This will ensure that both the tanker and loading equipment have the same electrical potential.
- ▶ Making ready fire control measures, such as fire fighting equipment, before transfer commences.
- ▶ Noting emergency escape routes before transfer commences.
- ▶ Closely monitoring the flow rates and fill capacities. This will include alarm systems to indicate when tanks are nearing their filling point.
- ▶ Having adequate venting arrangements in place to ensure vapour is dispersed properly and safely. This will include monitoring wind direction and strength.
- ▶ Low wind speed can be an added hazard, as the dispersion of vapour in these conditions is minimal and it can build up in dangerous quantities without being apparent. If there is an ignition source nearby, the results can be catastrophic.
- ▶ Having venting arrangements to the recipient tank in place as well as similar arrangements for the donor tank as air, or more likely inert gas, will have to replace the volume of product transferred.

Figure 14.7 Bottom loading of tanker
Source: Loadtec Engineered Systems Ltd.

Bottom loading of road tanker

Petroleum tanker filling via low level connections provides operators with the potential to load multiple compartments simultaneously as well as having automatic high level alarm for each compartment.

Tankers can also be top loaded, as can be seen in Figure 14.8.

Figure 14.8 Top loading of tanker
Source: Loadtec Engineered Systems Ltd.

14.6 Pressurized/refrigerated vessels for Liquefied Petroleum Gas (LPG), Liquefied Natural Gas (LNG) and carbon dioxide (CO_2)

Liquefied Petroleum Gas (LPG), Liquefied Natural Gas (LNG) and carbon dioxide (CO_2) all require specialized pressurized/refrigerated/insulated storage facilities in order to keep the product contained and safe. These tanks are likely to be cylindrical or spherical in shape as the laws of physics dictate that these shapes are the most efficient when it comes to containing high pressure gases.

Their construction is on a massive scale, with walls being up to 15 cm thick in order to contain the high pressure. However, there will be a pressure relief valve situated at the top of each tank which will be set to automatically operate if pressure builds up to a level near to the tank's maximum safe working limit.

We shall now look in more detail at individual requirements demanded by different products.

14.6.1 Liquefied Petroleum Gas (LPG) storage

As we mentioned in Chapter 2, to store Liquefied Petroleum Gas (LPG) effectively, it has to be converted from a gas to a liquid, which means it is stored at a temperature of between 0°C and −44°C. Apart from the containment of pressure within the tank, the other main safety issue is that any moisture which settles to the bottom of the tank will need to be drained off, and there is a risk that this water will freeze within the drain valve whilst it's in an open position, allowing LPG to escape.

This can be overcome with the use of a multi-valve drainage system together with an established procedure and sequence for opening and closing the valves to drain water. Finally, only fully trained and authorized personnel should be allowed to implement this procedure.

Liquefied Petroleum Gas (LPG) containers that are subjected to fire of sufficient duration and intensity can undergo a boiling liquid expanding vapour explosion (BLEVE). Consequently, LPG spheres should be protected by a fire water deluge system and this will have to be supplemented with an adequate fire water runoff system.

Figure 14.9 Spherical storage tanks
Source: Dreamstime.

Large, spherical liquefied petroleum gas containers may have wall thicknesses of up to 15 cm. Furthermore, they will ordinarily be equipped with an approved pressure relief valve.

14.6.2 **Liquefied Natural Gas (LNG) storage**

As we mentioned in Chapter 2, Liquefied Natural Gas (LNG) needs to be stored at −162°C in order to maintain its liquid state. This effectively reduces its volume 600-fold.

Storage tanks that are designed to store liquefied natural gas have double skins, with the inner skin containing the LNG and the outer skin containing insulation materials.

Liquefied natural gas is kept in its liquid state by maintaining very low temperatures. The temperature within the tank will remain constant if the pressure is kept constant. However, some of the liquefied natural gas will naturally change back to its gaseous state and

this will be released from the tank under an operation known as auto-refrigeration.

14.6.3 **Carbon dioxide (CO_2) storage**

Liquid carbon dioxide (CO_2) storage tanks usually have a storage capacity range from 3,000 kg to 50,000 kg. The vessels can be horizontal or vertically mounted. They have an inner steel tank cloaked by a layer of 175 mm insulation, and are covered with aluminium cladding.

Figure 14.10 Cylindrical storage tanks
Source: Dreamstime.

14.7 Loss of containment and consequences

Loss of containment is the one single event which has the greatest potential for catastrophic consequences, including loss of life and plant. Understanding what can happen to that product once it is released is therefore essential in implementing controls to mitigate the effects.

To this end, we will now be looking at the following issues:

▶ Jet fires and pool fires
▶ How hydrocarbon vapour clouds are generated and their potential consequences

14

▶ Boiling Liquid Expanding Vapour Explosions
(BLEVEs), Confined Vapour Cloud Explosions (CVCEs)
and Unconfined Vapour Cloud Explosions (UVCEs)
▶ Pipelines, their protection, surveying, maintenance,
security against arson and illegal tapping

14.7.1 Jet fires

A jet fire is where a fire emanates from a source of
fuel which is being continuously released in a particular
direction with significant force.

Jet fires can arise from a release of either a pure
flammable liquid or from a gaseous flammable liquid.

Figure 14.11 shows a typical jet fire.

Figure 14.11 Jet fire
Source: Dreamstime.

Jet fires are a significant element of the potential
consequences of a major offshore accident. The
characteristics of a jet fire mean it can engulf objects
with a high temperature flame almost instantly, which
can lead to structural failure of vessels or pipework and
a possible escalation of the situation.

Furthermore, because jet fires develop almost instantly,
they are difficult to control or isolate.

The properties of a jet fire depend on a number of
variables. These include the type of fuel, the rate of fuel
being released, the velocity of release, the direction of
release and the local wind direction and strength.

As we mentioned, jet fires reach full intensity almost
instantly. However, the contrary is also true in that they
can be turned off quickly if the source of fuel can be
terminated. This is an important characteristic when
control and isolation strategies have to be considered.

14.7.2 Pool fires

A pool fire is a fire which burns above a horizontal pool
of vaporizing hydrocarbon fuel.

Pool fires can be static, where the pool of fuel remains
in one position, or they can be running fires where the

pool of fuel runs like a river. These characteristics pose
a serious risk, especially where installations have large
liquid hydrocarbon inventories.

Figure 14.12 shows a typical pool fire.

Figure 14.12 Pool fire
Source: iStock.

14.7.3 How hydrocarbon vapour clouds are generated and their potential consequences

Any release of flammable liquids or gases can result in
the formation of a cloud of flammable vapour which,
as it is likely to be more dense than the surrounding
atmosphere, will stay at a low altitude as opposed to
rising and dispersing in the upper atmosphere. If the
cloud then meets a source of ignition, and it is within its
flammability limits, it will result in a fire or explosion.

Depending on the characteristics of the incident,
this may be a flash fire, a fire ball or a vapour cloud
explosion. If the vapour cloud still extends back to its
original source, it could result in a jet fire or pool fire.

Under certain conditions, the speed of the front of the
flame may move to a supersonic level, resulting in a
significantly more powerful explosion.

14.7.4 BLEVEs, CVCEs and UVCEs

The abbreviations listed below are all types of explosion
related to vapour clouds:

▶ BLEVE – Boiling Liquid Expanding Vapour Explosion
▶ CVCE – Confined Vapour Cloud Explosion
▶ UVCE – Unconfined Vapour Cloud Explosion

We will now look at what each of these situations
involves.

14.7.5 BLEVE – Boiling Liquid Expanding Vapour Explosion

Any vessel which is partly filled with pressurized
hydrocarbon liquid (e.g. liquefied petroleum gas) will
have a certain amount of space above it filled with
vapour. If the vessel is subjected to a fire, the pressure

Figure 14.13 BLEVE
Source: Wise Global Training.

in the tank will increase due to the liquid going above its boiling point and turning into a vapour.

The pressure relief valve will allow the overpressure to be vented to atmosphere in the first instance, but this will reduce the amount of liquid in the tank still further and the potential for the flame to engage with a section of the tank containing vapour and not liquid will increase. If this happens, the tank wall will weaken at this point as the heat transfer to vapour is much less efficient than to a liquid.

The result is likely to be a sudden and catastrophic failure of the vessel, with a discharge of vapour followed by an explosion when it reaches the flames. This, in effect, is a boiling liquid expanding vapour explosion.

14.7.6 Confined Vapour Cloud Explosion (CVCE)

A Confined Vapour Cloud Explosion (CVCE) is an explosion following a leak of vapour which occurs in a confined space, such as a building or a tank.

Where the confined space is totally confined, a vapour cloud explosion would generate approximately 8 bars of pressure. Most buildings would not be able to withstand this level of overpressure unless there was some venting. The venting could be a deliberate design feature (an explosion panel) or unintentional venting (a window or door giving way under the pressure).

A confined vapour cloud explosion inside a tank may cause the tank to rupture, with the potential for missile damage in the form of flying debris from the fragmented tank causing damage and injuries, and possibly fatalities.

14.7.7 Unconfined Vapour Cloud Explosion (UVCE)

An Unconfined Vapour Cloud Explosion (UVCE) is an explosion following a leak of vapour which occurs in an unconfined space (outdoors).

The actual process of what happens when an unconfined vapour cloud explosion takes place is as follows:

1 There is an initial and highly rapid expansion of burnt gas (fire ball).
2 This initial expansion of burnt gas (fire ball) produces a flow of unburnt gas ahead of it.
3 The flow of unburnt gas ahead of the flame front generates turbulence when obstacles are encountered.
4 This turbulence increases the flame area and with it, the burning velocity. This then increases the speed of the flame front as it passes through any obstacles. In effect, this increases the size of the explosion.
5 The increase in flame speed increases the volume of the fire ball as well as the flow of unburnt gas ahead of the flame.
6 At this point, we have an almost self-serving situation where the combustion-driven flow increases turbulence which increases the rate of combustion. Under these circumstances, it is quite possible for flame speeds to reach the order of the speed of sound.

An example of an unconfined vapour cloud explosion was the Flixborough disaster (see Figure 14.14).

Figure 14.14 Flixborough
Source: Getty Images.

14

14.7.8 Pipelines, their protection, surveying, maintenance, security against arson and illegal tapping

Oil and gas pipelines carry large amounts of product over vast distances. They are the most efficient means of distributing product and the safest, although if and when there is a breach in a pipeline, the quantity of product lost can be significant. Statistics show that accidents involving trucks carrying oil or gas are 3,000 times more likely to occur than with a pipeline, 38 times more likely by barge and 25 times more likely by rail.

Pipelines are manufactured from steel or plastic tubing which can vary, depending on the volume required to be transported, from 10 cm to 120 cm thick. They are generally buried underground at depths of between 1 and 2 metres. The momentum of the oil is maintained by pump stations situated at regular intervals along the pipeline's length. These locations can also serve for Remote Terminal Units (RTU) which feed back information on the pipeline's dynamics as well as service points for inserting and recovering Pipeline Inspection Gauges (PIGs).

Because of the high volumes of product involved, and once flow through a pipeline stops, it stops completely, pipelines need to be surveyed and maintained to the highest standards.

Surveys, inspections and maintenance activities can be carried out using pipeline inspection gauges. These devices are inserted into the pipeline at special insertion points and can perform various operations including inspections, cleaning and other maintenance operations. Some devices can even measure pipe wall thicknesses and the extent and location of any corrosion. They can perform all these operations without having to stop the flow of product in the pipeline.

The insertion and removal of PIGs can be a hazardous operation because of the high pressure of product in the pipeline. Consequently, strict safe working procedures need to be established and implemented at the PIG receiving and launching station.

PIG receiving and launching stations are generally ancillary loops in the pipeline which allow the PIG to be received in a trap or launched at a launching point. Figure 14.15 shows a drawing of a typical PIG receiving and launching station.

The following is an outline of the procedure for receiving and removing a PIG.

PIG removal

1 Once the PIG is detected as being in the receiving station by the PIG signalling device, the PIG receiver isolation valve and the bypass valve are closed.
2 The drain valve is opened to allow product in the now isolated receiving trap to depressurize and drain off.
3 The vent valve is opened to allow the PIG receiver pressure to be vented down. At the same time, nitrogen is introduced into the receiving station in order to purge the receiver of hydrocarbon vapours. This is a most important aspect, as air mixed with hydrocarbon vapours would create a potentially explosive situation.
4 Once the operator is satisfied that all product has drained out of the receiving station and it is fully purged, the receiver door can be opened and the PIG removed.

Now we have covered PIG removal, we can look at how a PIG is launched. Just like receiving and removing a PIG, launching can also be a hazardous operation. The following is an outline of the procedure for launching a PIG.

PIG launching

1 The kicker valve and the PIG launcher isolation valves are closed in order to isolate the launch station section.
2 The drain valve is opened to allow product in the now isolated launch station to depressurize and drain off.
3 The vent valve is opened to allow the PIG launch station pressure to be vented down. At the same time, nitrogen is introduced into the launch station in order to purge the station of hydrocarbon vapours.

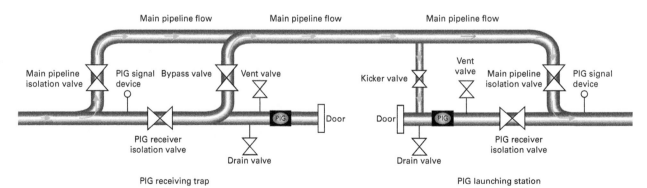

Figure 14.15 PIG receiving and launching station
Source: Wise Global Training.

4   Once the operator is satisfied that all product has drained out of the launch station and it is fully purged, the launch station door can be opened and the PIG inserted.

5   The launch station door is now closed and the station is now re-purged with nitrogen.

6   The kicker valve is opened to introduce product back into the pipeline.

7   The launch isolation valve is now opened, allowing the PIG to advance into the main pipeline.

Pipelines can be damaged by trailing anchors under the sea and trawling activities. Deliberate damage can come from arson or terrorist attacks or from illegal tapping.

Security devices are available which can monitor pipelines that are land-based or sub-aqua based. These work using fibre optics and can monitor both pipeline security and leak detection.

On land-based systems, they can monitor both underground and overground pipelines. As well as leaks, they will detect any activity being undertaken on the pipeline and relay its location back to a control room or security centre.

Sub-aqua systems are capable of detecting leaks and even diver activity within the vicinity of the pipeline.

Land-based pipelines which are situated underground should also have their location identified above ground with milestone markers. This is to help protect against damage from activities such as excavations, etc.

14.8 Decommissioning of plant and associated facilities

Decommissioning of an installation and its associated facilities involves planning, gaining approval for, and implementing the removal and disposal or reuse of the installation once it is no longer required for its current purpose.

There are five distinct stages in the decommissioning process:

1   All options for the physical dismantlement and/or removal of the installation are assessed and the best option is selected. This is then put through a detailed planning process involving all engineering, safety and environmental aspects.

2   Production/extraction is stopped and all the wells are plugged.

3   The installation is dismantled and/or removed in line with the agreed plan of action.

4   Those parts of the installation that have been removed are disposed of or recycled.

5   A seabed survey is carried out to ensure nothing untoward has happened to the original location of the installation. If any part of the installation remains, ongoing monitoring will be implemented.

There are strict international, regional and national guidelines and regulations regarding the decommissioning of oil and gas installations. These have to be taken into account at the very beginning of the decommissioning process when planning is under way. Most installations will either be reused at another location or recycled in some way.

Deciding on the best decommissioning process carries the highest degree of responsibility and care. It involves balancing the demands of providing an economic solution with protecting the environment, the health and safety of those involved in the decommissioning process, the technological issues involved, other users of the sea in the vicinity of the location and any regulatory or legal considerations.

The key to safe decommissioning is ensuring that rigorous and wide ranging risk assessments are conducted which encompass all the work involved in the decommissioning process. From this, safe systems of work will evolve and these should be observed and adhered to.

14.9 Management of simultaneous operations

Simultaneous Operations (SIMOPS) or activities can be described as a period in time when more than one operation or activity is being conducted at the same time with the potential that these activities could clash and bring about an undesired event or set of circumstances.

It is important that all work being undertaken at the same time, or which overlaps within a specific time frame, is identified before that work commences. It is then possible to review all activities and see where any conflict might occur.

It is not just the physical activities that need consideration, the demand for resources, or the termination of energy sources, may also cause conflict which could create unwanted situations.

Prior to any work commencing, the parties involved in the operations which are going to be conducted within a shared timeframe should each draw up a risk assessment of all their anticipated activities, including any constraints or hazards associated with those activities. They should then draw up a list of the corresponding measures which will mitigate those hazards that have been identified.

The parties should then meet to discuss the extent of their work and their requirements. The intention of the meeting is for each party to be able to draw up a work-specific dossier which covers how their activities will be conducted whilst taking into account all the other parties' activities and avoiding any conflict.

14

The meeting should resolve the following issues:

▶ The extent of the responsibilities of each party
▶ Nomination of a responsible person for each party
▶ Ensuring that responsible persons know of each other and have contact arrangements in place
▶ Identifying the requirements of each party
▶ Identifying the time frame of the actual work activity of each party

Compliance with procedures and good communications, including handover procedures, should be adhered to as standard.

Where permit-to-work systems are ongoing, they should take into account any conflict that simultaneous operations might create. They provide, amongst many other things, for the control of work activities that may interact or affect one another. They also provide a management facility where two or more individuals or groups need to co-ordinate activities to complete the job safely.

With regards to the permit-to-work system in place at the time of the Piper Alpha disaster, there was no cross-referencing when the work carried out under one permit might affect the work under another. Reliance was placed entirely on the memory of the designated authority. This practice has since changed so that permits-to-work do, as a matter of course, take into account other activities which may be affected.

REVISION QUESTIONS FOR ELEMENT 3 CONTINUED

9　With regard to fixed and floating roof storage tanks, **explain** what erosion is and where it might be expected to occur.

10　**Explain** what hazards are associated with fixed roof storage tanks.

11　With regard to floating roof storage tanks,
　(a)　**Explain** what 'landing the roof' is and what hazards are associated with it.
　(b)　**Explain** what control measures need to be taken if the roof is landed and why.

12　**Explain** what a 'bund' is and its purpose and function.

13　**Define** what a BLEVE is and **describe** an example of how a BLEVE might develop.

You can compare your answers with our guidance at the end of the book under the section entitled 'Revision and assessment guide'.

CHAPTER 15

Element 3

Sub-element 3.5:
Fire hazards, risks and controls

This chapter covers the following learning outcome

Outline the fire hazards, risks and controls relating to hydrocarbons.

15.1 Lightning

Figure 15.1 Lightning
Source: iStock.

A lightning strike is a massive discharge of electricity from the atmosphere, where the electrical charge has built up, to the earth.

The threat from lightning cannot be entirely eliminated, particularly with floating roof tanks where vapour is usually present around the rim seal. In these circumstances, measures to mitigate the consequences of a fire should be provided, including automatic rim seal fire extinguisher systems.

Threats from a lightning strike include:

▶ Sparks which can cause a fire or explosion
▶ Power surges to electrical equipment, particularly monitoring and safety devices which can render them inoperable

Protection from lightning strikes is a specialist area requiring expert knowledge as to what systems are suitable for each facility. However, in general they include the following:

▶ A 'dissipation array system' which reduces the potential between the site and any storm cloud cell that might be in the vicinity.
▶ A grounding system called a 'current collector'. This provides an electrically isolated area within which the facility will be located. This is normally made up of wire buried to a depth of about 25 centimetres and which surrounds the protected area. This wire is also connected to rods which are driven into the earth at about 10-metre intervals. Finally, the enclosed area is integrated by a net of cross-conductors which are also connected to any structures within the area, as well as the grounding system itself. This allows any current to discharge to earth safely.
▶ Electrical surge suppression devices. These devices have two distinct functions to perform. First, to stop direct strikes within the facility, and second, to prevent fast-rising, high current surges.

In general, the necessary precautions are:

▶ To keep the lightning channelled far away from the immediate neighbourhood of flammable and explosive materials
▶ To avoid sparking or flashover in joints and clamps, and at nearby components
▶ To prevent the overheating of conductors
▶ To prevent flashover or sparking due to induced voltages
▶ To prevent raising the potential of the earth termination system
▶ All metal containers to be of sufficient thickness (usually 5 mm minimum)
▶ Down-conductors fitted to all other metal structures and in sufficient numbers as to subdivide any current surge adequately
▶ All earthing systems to be interconnected to a single earth termination system. This usually takes the form of a mesh or grid pattern around the site.

15.2 Fire triangle and potential consequences of explosions and thermal radiation

Fire can be defined as 'the rapid oxidization of a material or substance'. This is known as combustion, which releases light, heat and various reaction products such as smoke and gas.

Fire is made up of three interdependent elements known as the fire triangle. These are:

▶ Heat or a source of ignition
▶ Fuel
▶ Oxygen

This is known as the fire triangle.

Figure 15.2 Fire triangle
Source: Wise Global Training.

15.2.1 The fire triangle

The fire triangle is a way of understanding the way in which these elements, which are necessary for most fires to burn, interrelate.

The triangle shows that for burning to start (and to continue to do so) requires three elements: heat, fuel and an oxidizing agent (this is usually oxygen – but not always).

A fire naturally occurs when these elements are brought together in the right proportions.

A fire can be prevented or extinguished by removing any one of these elements.

Similarly, fires can be prevented by the isolation of any one of these elements, particularly fuel and ignition.

It is important to remember that it is only the vapour from a fuel that actually burns. Before combustion takes place, a solid or liquid must be heated to a point at which vapour is given off and that vapour can ignite.

15.2.2 Explosions

An explosion is a type of fire but one which combusts with such a rapid force that it causes an effect known as over-pressure (explosion). Under certain conditions, the speed of the front of the flame may move to a supersonic level, resulting in a significantly more powerful explosion.

There are three types of explosion that are associated with the oil and gas industry. These are:

1 **Boiling Liquid Expanding Vapour Explosion (BLEVE)**
2 **Confined Vapour Cloud Explosion (CVCE)**
3 **Unconfined Vapour Cloud Explosion (UVCE)**

We touched on these types of explosion in Chapter 14 but, because of their significance in the oil and gas industry, it's worthwhile reminding ourselves of what these types of explosion are.

15.2.3 Boiling Liquid Expanding Vapour Explosion (BLEVE)

We will now look at a scenario whereby a Liquefied Petroleum Gas (LPG) storage vessel is exposed to an external source of heat – possibly a fire. The LPG is kept in a liquid state inside the vessel because it is held at a temperature below its boiling point, which means it takes up substantially less space than in its gaseous state. Anything which changes that state from a liquid to a gas, such as the external source of heat (fire), will increase the pressure inside the storage vessel to a potentially unsustainable level.

At first the vapour will be vented out via the pressure relief valve on the top of the vessel. However, the rate of increase in pressure under these circumstances is likely to be unsustainable and the vessel is likely to eventually fail, with a consequential loss of containment. The resulting instantaneous release of LPG vapour will likely make contact with a source of ignition, resulting in a Boiling Liquid Expanding Vapour Explosion (BLEVE).

Should a situation occur whereby a source of heat (fire or otherwise) begins to radiate itself onto an LPG storage vessel, the following action should be taken. Apart from removing or extinguishing the source of heat, the storage vessel, and any other storage vessels nearby, should be deluged with copious amounts of water to keep the metal cool.

15.2.4 Confined Vapour Cloud Explosion (CVCE)

A confined vapour cloud explosion is an explosion following a leak of vapour which occurs in a confined space, such as a building or a tank.

15.2.5 Unconfined Vapour Cloud Explosion (UVCE)

An unconfined vapour cloud explosion is an explosion following a leak of vapour which occurs in an unconfined space, outdoors.

15.2.6 Thermal radiation

Thermal radiation is the transfer of heat from one source to another. This can be a structure or a person. Where the recipient source is a person, the consequences can be severe.

The initial effect of exposure to a source of heat (fire) is to warm the skin. This then becomes painful as the amount of energy absorbed increases. Thereafter, second-degree burns begin to take effect, with the depth of burn increasing with time for a steady level of radiation. Ultimately, the full thickness of the skin will burn and the underlying flesh will start to be damaged, resulting in third-degree burns.

When plant, including pipework and vessels, is exposed to thermal radiation the effect is the transfer of heat to the product inside the plant. This can change the characteristic of the product and make it less stable. These characteristics include the potential to make the product expand and/or increase the amount of vapour given off, amongst other things. This can result in loss of containment, with an ensuing vapour cloud explosion, jet fire, pool fire or running liquid fire.

15.3 Electrostatic charges

Whenever a liquid moves against a solid object, such as the inside of a pipe, it generates a static electrical charge. This is caused by ions (charged atoms) being transferred from the liquid to the surface of the pipe or vessel.

15

The most common cause of static electricity build-up is where there is a flow (transfer) or movement (mixing process) of liquid within a process.

The amount and rate of static generation can be dictated by a number of factors. These factors, or their elimination or reduction, can also be used to control the risks associated with static electrical generation. These include:

▶ The conductivity of the liquid
▶ The amount of turbulence in the liquid
▶ The amount of surface area contact between the liquid and other surfaces
▶ The velocity of the liquid
▶ The presence of impurities in the liquid
▶ The atmospheric conditions. Static build-up is enhanced when the air is dry.

Let's look at some typical areas within a process where static electricity is most likely to occur, as well as some simple control measures.

Electrostatic charges – piping systems

As we've mentioned, the flow of liquid through piping systems can generate a static charge. However, there are factors which can influence the amount of charge generated. These include the rate of flow and the velocity of the liquid.

Control measures include keeping the rate and velocity of the liquid low. This can be achieved by ensuring pipe dimensions are appropriate for the volume of liquid flowing through them; and also ensuring the length of pipe is as short as possible.

Electrostatic charges – filling operations

Filling operations, which involve large flows of liquid and splashing, generate turbulence. This turbulence allows the large amounts of liquid to pass against the vessel surfaces which in turn generates a static charge. If the liquid has already passed through piping to get to the filling operation, this will only serve to increase the accumulated charge already generated.

Control measures include:

▶ Ensuring filling operations do not involve the free-fall of liquids. This will reduce the amount of splashing taking place.
▶ Lowering the velocity of the liquid being filled.
▶ Ensure fill pipes touch the bottom of the container being filled.
▶ Tanks which have been filled with products that have a low conductivity, i.e. jet fuels and diesels, should be given time to relax before the process continues.
▶ Tanks which have been filled with product should not have any ullage (vapour space) for a set period of time. Nor should any dipping of the product take place, again for a set period of time.

Electrostatic charges – filtration

By their very nature, filters have large surface areas, and this can generate as much as 200 times the amount of electrostatic charge in a piping system that has a filtration system within it, as compared with the same piping system without filtration.

Control measures include ensuring good bonding and grounding is in place (see below).

Electrostatic charges – other issues

▶ Liquids which have particles within them are more susceptible to the generation of static charge than those without.
▶ Static can be generated when liquids are mixed together.
▶ Piping or vessels which allow a space for vapour to accumulate are a particular concern as any spark generated from a discharge of static electricity may cause an explosion inside the pipe.

15.3.1 Methods of controlling static charges

Although the generation of static electricity cannot be totally eliminated, the rate of generation and its accumulation can be reduced by the following control measures:

Methods of controlling static charges – additives

In some instances, anti-static additives can be introduced to reduce static charge build up.

Methods of controlling static charges – bonding and grounding

Bonding and grounding techniques are a very effective means of minimizing the risk of spark generation from a build-up of static electricity.

A bonding system is where all the various pieces of equipment within a process system are connected together. This ensures that they all have the same electrical potential, which means there is no possibility of a discharge of electricity, by way of a spark, from one piece of equipment to another.

Grounding is where pieces of equipment (which may be bonded together or not) are connected to an earthing point. This ensures any electrical charge in the equipment is given the means to constantly flow to earth, thus ensuring there is no potentially dangerous build-up of charge which could lead to a sudden discharge of electricity, by way of a spark.

All equipment which is involved in processing or storing flammable liquid, gas or vapour should be bonded and grounded.

Some other considerations are:

▶ Incidental objects and equipment, such as probes, thermometers and spray nozzles, which are isolated, but which can become sufficiently charged to cause a static spark, may need special consideration.

▶ The cables used for bonding and grounding cables should be heavy duty cables. This is to ensure that they can cope with physical wear and tear without compromising their grounding ability. It is also to ensure that their electrical resistance is as low as possible.

▶ The bonding of process equipment to conductors must be direct and positive.

▶ Using an inert gas, such as nitrogen, within the ullage space of a storage vessel will prevent an explosion or flash fire occurring if an electrostatic spark does occur. The inert gas lowers the oxygen content of the gas in the ullage space, thus ensuring there is insufficient oxygen to support a burning process (oxygen being part of the fire triangle).

▶ Operators should wear anti-static clothing.

15.4 The identification of ignition sources

15.4.1 Fire hazards, risks and controls

In the oil and gas industry, the severity of any incident involving fire and/or explosion is likely to be very grave, possibly involving loss of life, severe damage or destruction of plant, as well as having a potential impact on local communities. Consequently, any type of fire or explosion is unacceptable and controls must be put in place to prevent such an occurrence. These controls fall into two main categories.

First, any product should remain contained or under control throughout the process it is undergoing. In simple terms this means that any leak of product is regarded as highly undesirable. However, if a leak does occur there should be systems in place to detect it immediately and for appropriate action to be taken to control it and/or mitigate any consequences.

Second, all sources of ignition should be eradicated as far as possible in areas where product is processed and has the potential to escape.

Where it is necessary to introduce an ignition source into such an area, such as maintenance involving hot work, then an appropriate risk assessment should be undertaken to identify and evaluate the risks, as well as introducing a permit-to-work regime. These measures may well be accompanied by other appropriate controls, such as temporarily shutting down the process and having fire-fighting equipment to hand.

15.4.2 Identifying sources of ignition

We will now look at potential ignition sources which need to be considered when conducting a risk assessment. Some of the sources of ignition have had basic control measures added.

▶ Smoking and smoking material
  ▷ A total ban on smoking and the taking of smoking materials into controlled areas should be enforced.
▶ Vehicles
  ▷ Vehicles may be totally prohibited or restricted to only specially adapted vehicles.
▶ Hot work such as welding, grinding, burning, etc.
  ▷ Implement a permit-to-work regime.
▶ Electrical equipment
  ▷ The equipment should be suitable for the zone it is intended to be used in. It should also be properly and regularly inspected and maintained.
▶ Machinery such as generators, compressors, etc.
▶ Hot surfaces such as those heated by process or by local weather (hot deserts)
▶ Heated process equipment such as dryers and furnaces
▶ Flames such as pilot lights
▶ Space heating equipment
▶ Sparks from lights and switches
  ▷ Use only electrical equipment and instrumentation classified for the zone in which it is located.
▶ Impact sparks
▶ Stray current from electrical equipment
  ▷ Ensure all equipment is bonded and earthed.
▶ Electrostatic discharge sparks
  ▷ Bond and ground all plant and equipment.
▶ Electromagnetic radiation
  ▷ Make the correct selection of equipment to avoid high intensity electromagnetic radiation sources, e.g. limitation on the power input to fibre optic systems, avoidance of high intensity lasers or sources of infrared radiation.
▶ Lightning
  ▷ We have covered the control measures for lightning earlier in this section. There should be measures in place which reduce the potential of a lightning strike, as well as a grounding system to disperse any charge that may affect the installation. A further consideration is to look at weather windows (i.e. to not work during electrical storms).

Other control measures include:

▶ Controls over activities that create intermittent hazardous areas, e.g. tanker loading/unloading.
▶ Control of maintenance activities that may cause sparks or flames through a permit-to-work system.
▶ Precautions to control the risk from pyrophoric scale. This is where a substance can ignite spontaneously

15

in air, particularly humid air, and is usually associated with formation of ferrous sulphide.

▶ Where control and/or detection equipment is regarded as critical, such as smoke and flame detectors, then a back-up or secondary system may be considered appropriate.

All of these control measures are supplementary to the main control and fire-fighting systems such as emergency shutdown systems, fire deluge systems, sprinkler systems, etc.

15.5 Zoning/hazardous area classification and selection of suitable ignition-protected electrical and mechanical equipment and critical control equipment

15.5.1 Introduction

Gases and vapours can create explosive atmospheres. Consequently, areas where these potentially hazardous airborne substances present themselves are classed as hazardous areas so that appropriate controls can be implemented.

However, how often these substances present themselves is also a factor in determining the appropriate level of control. For example, if the presence of a flammable vapour only happens once every three months, it would not be sensible to apply the same level of control to an area where a flammable vapour is present all day, every day.

The answer is to apply a classification to areas – called zoning – which places appropriate controls on the type of equipment that can be used in that area and which potentially can create a source of ignition, particularly electrical equipment, which reflect the risk involved.

This zoning is determined by the frequency and extent of explosive atmospheres being present over a fixed period of time and the likelihood of an explosive atmosphere occurring at the same time as an ignition source becomes active. All of these parameters are established through a rigorous risk assessment.

15.5.2 Zoning

A place where an explosive atmosphere may occur on a basis frequent enough to be regarded as requiring special precautions to reduce the risk of a fire or explosion to an acceptable level is called a 'hazardous place'.

A place where an explosive atmosphere is not expected to occur on a basis frequent enough to be regarded as requiring special precautions is called a 'non-hazardous place'.

Under these circumstances, 'special precautions' means applying measures to control sources of ignition within an area designated as a hazardous place.

Determining which areas are hazardous places, and to what extent, is called a 'hazardous area classification study'. A hazardous area classification study is a method of analysing the extent and frequency to which an area is subject to having an explosive atmosphere. The main purpose of this is to facilitate the appropriate selection and installation of apparatus, tools and equipment which can be used safely within the environment, even if an explosive atmosphere is present.

A hazardous area classification study involves giving due consideration to the following:

▶ The flammable materials that may be present
▶ The physical properties and characteristics of each of the flammable materials
▶ The source of potential releases and how they can form explosive atmospheres
▶ Prevailing operating temperatures and pressures
▶ Presence, degree and availability of ventilation (forced and natural)
▶ Dispersion of released vapours to below flammable limits
▶ The probability of each release scenario

Consideration of these factors will enable the appropriate selection of zone classification for each area regarded as hazardous, as well as the geographical extent of each zone. The results of this work should be documented in hazardous area classification data sheets. These sheets should be supported by appropriate reference drawings which will show the extent of the zones around various plant items.

Hazardous areas are classified into zones based on an assessment of two factors:

1 The frequency of the occurrence of an explosive gas atmosphere
2 The duration of an explosive gas atmosphere

These two factors in combination will then facilitate the decision-making process which will determine which zone will apply to the area under consideration.

▶ **Zone 0**: An area in which an explosive gas atmosphere is present continuously or for long periods of time
▶ **Zone 1**: An area in which an explosive gas atmosphere is likely to occur in normal operation
▶ **Zone 2**: An area in which an explosive gas atmosphere is not likely to occur in normal operation but, if it does occur, will only exist for a short period of time

As the zone definitions only take into account the frequency and duration of explosive atmospheres being present, and not the consequences of an explosion, it may be deemed necessary, because of the severe consequences of any explosion, to upgrade any equipment specified for use within that area to a higher level. This will be a discretionary option open to the analysis team.

15.5.3 Selection of equipment

As we inferred earlier in this section, the whole idea of zoning is to determine what apparatus, tools and equipment may be installed or used in a particular zone. The issue with electrical equipment is that it normally creates sparks, either as a result of the brushes coming in contact within the rotating armature, or when a switch is activated. Either event can ignite any flammable gas present in the atmosphere in the vicinity of the equipment.

Consequently, manufacturers have designed specialized equipment which overcomes, in various ways, the issue of having sparks which are exposed to the local atmosphere. The particular solution which is incorporated into each piece of equipment is signified by a code which is marked on the equipment's product identification label. For example, 'd' signifies equipment which has the motor and switch enclosed in a flameproof enclosure, or 'q' powder filled. Both

pieces of equipment are safe to use in zones 1 and 2, as indicated in Table 15.1 below.

The guide to what equipment is appropriate for each zone is the ATEX equipment directive and, whilst this is not a legal requirement outside the EU, most of the electrical standards have been developed over many years and are now set at international level.

Apparatus, tools and equipment are categorized in accordance with their ability to meet the standards required when used within each zone, as shown in Table 15.1.

As well as taking into account the sparks that electrical equipment can generate, consideration also needs to be given to the potential surface temperature of all equipment, not just electrical equipment, although most electrical equipment does generate heat as a matter of course.

In order to facilitate this, temperatures have been categorized into six classes: T1–T6. The bigger the T-number, the lower the allowable temperature of any equipment used. The temperature class will be

Table 15.1 Tools and equipment categorization in zoned areas

Zone 0	Zone 1	Zone 2
An area in which an explosive gas atmosphere is present continuously or for long periods of time	An area in which an explosive gas atmosphere is likely to occur in normal operation	An area in which an explosive gas atmosphere is not likely to occur in normal operation but, if it does occur, will only exist for a short period of time
Category 1 equipment Note: Although this equipment is categorized for use in Zone 0, it can also be used in Zones 1 and 2	Category 2 equipment Note: Although this equipment is categorized for use in Zone 1, it can also be used in Zone 2	Category 3 equipment Note: This equipment can only be used in Zone 2
'ia' – Intrinsically safe Ex s – Special protection if specifically certified for Zone 0	'd' – Flameproof enclosure 'p' – Pressurized 'q' – Powder filled 'o' – Oil immersion 'e' – Intrinsically safe 'ib' – Intrinsically safe 'm' – Encapsulated 's' – Special protection	Electrical type 'n'

Source: Adapted from 'Hazardous Area Classification and Control of Ignition Sources' available at www.hse.gov.uk/comah/sragtech/techmeasareaclas.htm

Table 15.2 Temperature classification for tools and equipment in zoned areas

Temperature classification	Maximum surface temperature	Substances can be used which will not auto ignite at temperatures below
T1	450°C	450°C
T2	300°C	300°C
T3	200°C	200°C
T4	135°C	135°C
T5	100°C	100°C
T6	85°C	85°C

Source: Adapted from 'Hazardous Area Classification and Control of Ignition Sources' available at www.hse.gov.uk/comah/sragtech/techmeasareaclas.htm

15

133

determined by the auto ignition temperature of the substance involved (see Table 15.2).

The T-number of each piece of equipment will also be marked on the equipment's product identification label.

If several different flammable materials may be present within a particular area, the material that gives the highest classification will dictate the overall area classification of any equipment used.

CHAPTER 16

Sub-element 3.6:
Furnace and boiler operations

This chapter covers the following learning outcome

Outline the hazards, risks and controls available for operating boilers and furnaces.

Chapter contents

16.1 Use of furnace and boiler operations

16.1.1 Introduction

Boilers are devices which heat large quantities of water in order provide a constant supply of hot water, or to turn it into steam. Where steam is generated, this is captured and kept in a pressurized state.

Furnaces, or **process heaters**, are devices which are used to provide a large source of heat to various process streams and are used extensively in the oil and gas industry. Examples of their use include process

Figure 16.1 Boilers
Source: Dreamstime.

heating where product is brought up to a required temperature; crude oil heating prior to it being separated into its various component parts (fractional distillation); and cracking, which converts product with large molecules (generally low value product) into product with smaller molecules (generally higher value product, i.e. petrol).

16.1.2 Boiler/furnace components

- ▶ The combustion chamber is where the heat is generated by burning the fuel.
- ▶ The heat exchanger is where the heat is transferred to water (or product if it is a furnace/process heater).
- ▶ The chimney or flue allows exhaust gases to escape to atmosphere. Some boilers/furnaces also have a heating coil positioned within the flue. This extracts some of the heat from the exhaust gases which would otherwise be lost to the atmosphere. These devices are known as waste heat heaters or economizers.
- ▶ The controls, of which there are a number, allow the heating of water or product to be done in a regulated, efficient and safe manner. The combustion and operating controls regulate the rate of fuel used to meet the demand.

16.1.3 Boiler/furnace safety components

Over and above the main components, all boilers used for steam production have a safety relief valve which allows excessive steam pressure to be released to prevent overpressure or explosion.

Boilers also have a drain which allows sediments and contaminates to be drained from the water. This can be a continuous process or one which is undertaken periodically. This process is known as 'blow down' and is dealt with in detail later in this chapter.

Other boiler/furnace safety controls include automatic and continuous monitoring of pressure and temperature, high and low gas or oil pressure, high and low water/product levels and flame safeguard controls. Generally, these are connected to a circuit breaker which prevents the firing of the boiler if any of these safety controls are activated.

16.1.4 Boiler types

There are two main types of boiler: fire tube boilers and water tube boilers. As their names indicate, one has fire within its tubes, whereas the other has water within its tubes.

Let's now take a look at the differences as well as the advantages and disadvantages of both types of boiler.

Boiler types – fire tube boilers

A fire tube boiler, as shown in Figure 16.2, is one which generates hot gases which then pass through a number of tubes before being expelled out of the flue. These tubes run through a sealed and insulated container of water and the heat from the gases is transferred by thermal conductivity to the water, which then turns to steam. The steam from the boiler then exits through a tube at the top of the container.

Fire tube boilers are relatively inexpensive and are easy to clean. They are usually smaller than water tube boilers and they are easy to re-tube.

The disadvantages of fire tube boilers are they are not suitable for high pressure application above 1.7 MPa and are unable to generate high capacity steam.

Note: MPa stands for Mega Pascal and is a measurement of pressure. 1 MPa is equal to 9.87 bar or 145 psi.

Boiler types – water tube boilers

A water tube boiler, as shown in Figure 16.3, has basically the opposite configuration of fire tubes. In a water tube boiler, a number of tubes run through the furnace part of the boiler. This heats the water inside the tubes which turns it to steam. Because only

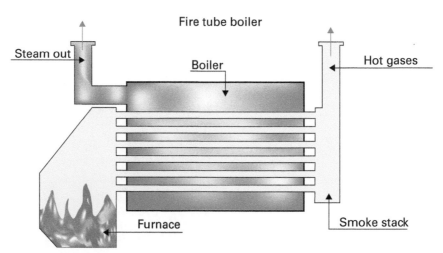

Figure 16.2 Section through fire tube boiler
Source: Wise Global Training.

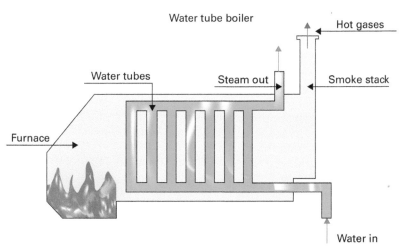

Figure 16.3 Section through water tube boiler
Source: Wise Global Training.

relatively small amounts of water are being heated at any one time, water tube boilers can generate steam to an extremely high temperature. Although on the downside, they can't generate as much steam as a fire tube boiler.

Water tube boilers are available in larger sizes than fire tube boilers, they are able to produce much higher pressures of steam (up to 34.5 MPa), recover faster than a fire tube boiler and can reach very high temperatures (superheat).

The disadvantages of water tube boilers are that they are an expensive capital outlay, they are more difficult to clean and are much bigger in size.

16.1.5 Furnace types

There are two main types of furnace/process heater: natural draught and balanced draught.

Furnace types – natural draught

This type of furnace/process heater uses the principle of a natural draught to move the air and combustion gases through the combustion chamber. It has a high chimney stack, which creates a difference in pressure between the cold air at the burner and the hot and less dense air at the top of the chimney stack. This creates a natural flow of combusted air up the chimney.

Furnace types – balanced draught

This type of furnace/process heater has a fan or blower to increase and control the flow of air and combustion gases through the combustion chamber. These types of furnace/process heater usually have a heat recovery system, known as a waste heat heater or economizer, fitted to the waste gas outlet stack. This extracts some of the heat from the exhaust gases which would otherwise be lost to the atmosphere, but the exhaust gases must be controlled in order to work effectively, which is what the operator can do via the fan/blower controls.

The main advantage of balanced draught furnaces/process heaters is that the increased air flow improves the heat capacity of the furnace/process heater. With this type of furnace/process heater, the height of the chimney stack is also less critical.

16.1.6 Furnaces

Let's now take a look at a typical furnace application. Furnaces are used extensively for fractional distillation. This is the process of separating a mixture into its component parts, for example crude oil into its useful components. The product is heated within the furnace and allowed to separate into its component parts at various temperatures, as shown in Figure 16.4.

16.2 Hazards and risks of operating boilers and furnaces, in particular those arising from loss of pilot gas supply, over-filling, flame impingement, firebox overpressure, low tube flow or control of Tube Metal Temperature (TMT)

16.2.1 Boiler/furnace hazards

There are a number of specific hazards associated with operating boilers and furnaces which need special attention. These include:

▷ Pilot lights
▷ Boiler over-firing
▷ Flame impingement
▷ Firebox overpressure
▷ Low tube flow
▷ Control of Tube Material Temperature (TMT)

We'll now take a more detailed look at each of these issues.

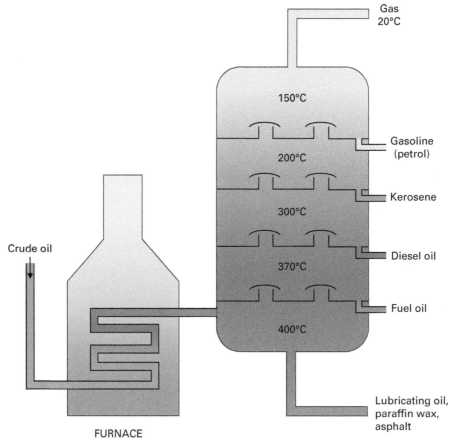

Figure 16.4 Fractional distillation process
Source: Wise Global Training.

Boiler/furnace hazards – pilot lights

Pilot lights provide a source of ignition to the main boiler when it needs to be fired up.

The gas supply to the pilot light and the main burner is controlled by the main gas valve. In a situation where the pilot light fails to light or goes out, gas will continue to enter the chamber, causing a build-up of flammable gas which, if ignited, could cause an explosion.

In order to counteract this scenario, a sensory device called a 'thermocouple' is located in close proximity to the pilot light in order to detect if there is heat coming from the pilot light. If no heat is detected, the device will activate a relay which will close the main gas valve.

Boiler/furnace hazards – over-firing

Boilers are designed to work within a specific heat flux range. Over-firing the boiler is basically allowing the heat flux to increase to a level beyond its upper Maximum Continuous Rating (MCR), which is set by the manufacturer of the boiler. This can then have an impact on, amongst other things, the furnace walls and the surface temperature of the refractory. It can also result in a substantial increase in tube and membrane operating temperatures, which can lead to a degradation of tube metallurgy and strength.

Boiler/furnace hazards – flame impingement

Flame impingement is where the flame produced by the burners comes into contact with the surface being heated. This is something which the design of the boiler or furnace does not normally allow. This is because the temperature of the surface being heated should always remain within prescribed parameters, and direct flame contact can lead to these parameters being exceeded.

If this occurs there tends to be a gradual build-up of carbon on the inside of the tube at the point where the flame is in contact. This layer of carbon acts as an insulator between the water/product inside the tube and the surface where the flame is impinging on it, compounding the effect the flame is having on the heated surface of the tube.

If this process is allowed to continue, it can lead to the tube eventually becoming blocked, resulting in the potential rupture of the tube.

Causes of flame impingement can be:

▶ Improper burner adjustment
▶ Poor forced draught conditions
▶ Poor design

Control measures

The first step is to ensure the flame is kept off the heated surface. Regular inspections in order to monitor flame behaviour are essential. These should be recorded in the boiler/furnace log.

If the flame is seen to impinge on the heated surface then the burner should be adjusted accordingly or the boiler/furnace shut down so that maintenance activities can determine the cause and remedy the fault.

Boiler/furnace hazards – firebox overpressure

Fireboxes are that part of a boiler or furnace where the heat is generated by the burning of fuel. Firebox overpressure typically occurs after a flameout, which happens when the burner flame is extinguished for some reason. The fumes, gases or vapour from the fuel will begin to build up inside the combustion chamber, which will invariably be hot. This will make them highly volatile, and when they reach their Lower Explosive Limit (LEL) and make contact with a source of ignition, this will cause overpressure (explosion).

Another form of minor firebox overpressure is when the combustion of the fuel is proceeding through a rapid series of detonations. This can be caused by an incorrect mixture of air and fuel.

Fireboxes normally have overpressure monitors, but if these malfunction or are incorrectly set, any onset of problems relating to potential firebox overpressure may go undetected.

Overpressure monitors should be tested regularly. Exhaust blowers should be inspected and maintained regularly. Flame quality and consistency should be regularly monitored.

Boiler/furnace hazards – low tube flow

Normal circulation within water or steam pipes heated by a boiler is generated by the difference in density between cooled water and hot water/steam. Cooled water is denser than hot water/steam and consequently gravitates downwards, resulting in the hot water/steam being pushed upwards. This principle provides a large circulation force.

Any adverse conditions within the flow system will affect this flow rate and create a situation where flow rates are eventually reduced. When this becomes acute, it is known as 'low tube flow'.

The reason for low tube flow can be partial choking of the tube's internal diameter, and this is generally due to a build-up of deposits from impurities within the water or from carbon build-up caused by flame impingement, which we covered earlier.

Low tube flow can also affect furnaces/process heaters, which can lead to increases in high tube metal temperatures and the subsequent carbonizing of product and tubes becoming restricted. Many furnaces/process heaters have tube pass flow meters fitted that

allow the operators to correctly balance the flow of product through the tubes.

Control measures include:

▶ Flushing the tubes regularly to ensure any blockages are removed
▶ With regard to boilers, keeping the water quality at the recommended level
▶ Ensuring flame impingement does not occur
▶ With regard to furnaces/process heaters, using tube pass flow meters to monitor product flow

Boiler/furnace hazards – Tube Material Temperature (TMT)

We've touched on a number of reasons for heated surfaces going beyond their prescribed working parameters, including boiler/furnace over-firing and flame impingement. We also mentioned that flame impingement causes scale to build up on the water side of heat exchange surfaces, creating a situation where the tube material temperature will rise.

This scenario can also be created where scale is allowed to build up, which can also lead to localized overheating and potential component failure. This is generally due to poor water quality, where suspended material tends to congregate in the bottom of the boiler and cause scale to develop. It can also be caused by ingress of product (e.g. oil) into the condensate return system.

Controls include regular testing of water and blowing down on a regular basis, as well as checking for leaks in the water/steam circulation system.

16.2.2 Maintenance

Now we've completed our look at hazards associated with operating boilers and furnaces, we need to look at what maintenance activities are required.

Maintenance – boiler blow down

Total Dissolved Solids (TDS) are substances such as minerals, salts and even metals, which are held in a suspended form within water. If these solids are of a sufficient concentration within the water used in a boiler system, they can attach themselves to the inside of boilers and, over time, build up to form scale.

Consequently, the first action is to maintain the solids below a certain limit. This is done by testing the water with a Total Dissolved Solids (TDS) meter or conductivity meter. This measures the conductivity of the water, which is an indication of the measure of TDS within the water. As the TDS concentration increases, the likelihood that the dissolved solids will precipitate out of the water and form scale also increases. At this point it is necessary to drain some of the water from the system – this is called boiler blow down – in order to remove some of those dissolved solids and keep the TDS concentration below the level where they will precipitate.

Large steam boilers should have a means of continuous blow down rather than periodic blow downs. Continuous blow down is where a small amount of water is continuously drained from the system whilst fresh water is introduced to make up the volume.

Maintenance – logs and boiler/furnace checklists

Maintenance is a critical factor in maintaining a safe and efficient boiler or furnace. A log should be kept of all inspections and maintenance activities, and inspections should be carried out against a checklist to ensure the equipment is operating properly.

Pressure, water/product temperature and flue gas temperatures should be recorded daily. These will serve as a baseline reference for detecting variations which could indicate developing problems.

More detailed inspections of a system's performance should be conducted and recorded on a regular basis. This will help to show trends in a system's operating condition, which may be very gradual and not readily apparent without the use of such documentation.

A daily checklist might comprise of the following:

- Boiler/furnace use and sequencing
- Overall visual inspection
- Lubricate all oil and grease nipples
- For boilers, check steam pressure
- For boilers, check water level
- Check burner
- Check motor
- Check air temperature
- Check oil filter
- Inspect oil heaters
- For boilers, check water treatment
- Complete log

There will be other checklists for activities which need conducting on a weekly, monthly and annual basis.

REVISION QUESTIONS FOR ELEMENT 3 CONTINUED

16 There are two main types of boiler used to heat water. **Explain** what they are and how both types work.

17 In relation to boilers and furnaces, **explain** what flame impingement is and the potential consequences if it is not rectified.

18 **Explain** what total dissolved solids are and how they can be controlled.

You can compare your answers with our guidance at the end of the book under the section entitled 'Revision and Assessment Guide'.

Sources of reference for Element 3

Bader, A., Baukal, C. E. Jr. and Bussman, W. (2011) 'Selecting the Proper Flare System'. AIChE. Available at: http://people.clarkson.edu/~wwilcox/Design/FlareSel.pdf

Bemment, R. for Health and Safety Executive (2001) *Offshore Technology Report 2001/032 Decommissioning Topic Strategy*. London: HSE Books. ISBN 0-7176-2054-9. Available at: www.hse.gov.uk/research/otopdf/2001/oto01032.pdf

Bright Hub Engineering (n.d.) 'How Long Term Overheating Tube Failures Happen in Boilers'. Available at: www.brighthubengineering.com/power-plants/38111-how-long-term-overheating-tube-failures-happen-in-boilers/

Buncefield Standards Task Group (BSTG) (2007) *Final Report: Safety and Environmental Standards for Fuel Storage Sites*. Available at: www.hse.gov.uk/comah/buncefield/bstgfinalreport.pdf

Cooper Crouse-Hinds (2009) 'AN9003 – A Users Guide to Intrinsic Safety'. Available at: www.mtl-inst.com/images/uploads/datasheets/App_Notes/AN9003.pdf

Endress + Hauser Messtechnik GmbH (2004) 'Functional safety in the Process Industry – Risk Reduction with Safety Instrumented Systems: Safety Integrity Level – Safety Instrumented Systems'.

Energy Institute (2008a) *Guidance for Corrosion Management in Oil and Gas Production and Processing*. London: Energy Institute. ISBN 978-0-85293-497-5.

Energy Institute (2008b) *Corrosion Threats Handbook: Upstream Oil and Gas Production Plant*. London: Energy Institute. ISBN 978-0-85293-496-8.

Funderburk, R. S. (1997) 'Key Concepts in Welding Engineering'. *Welding Innovation* vol. XIV, no. 2. Available at: www.jflf.org/pdfs/papers/keyconcepts1.pdf

Giffen, P. (2006) 'Safety Integrity Level'. Tech News, South African Instrumentation and Control. Available at: www.instrumentation.co.za/news.aspx?pklnewsid=21257

GRISHMA Global Technologies (n.d.) 'FZC Floating Roof Tank (rim seal) Fire Protection System'. Available at: www.grishmaglobal.com/download/RimSeal.pdf

Hardie, F. (2009) *Evaluation of the Effectiveness of Non-Destructive Testing Screening Methods for In-Service Inspection (RR659)*. London: HSE Books. Available at: www.hse.gov.uk/research/rrpdf/rr659.pdf

Health and Safety Executive (n.d.a) 'Atmospheres – Classification of Hazardous Areas (Zoning) and Selection of Equipment'. Available at: www.hse.gov.uk/fireandexplosion/zoning.pdf

Health and Safety Executive (n.d.b) 'Hazardous Area Classification and Control of Ignition Sources'.

Available at: www.hse.gov.uk/comah/sragtech/
techmeasareaclas.htm

Health and Safety Executive (n.d.c) 'Human Factors:
Safety Critical Communications'. HSE. Available at:
www.hse.gov.uk/humanfactors/topics/common3.pdf

Health and Safety Executive (n.d.d) 'Integrity of
Atmospheric Storage Tanks'. Available at: www.hse.
gov.uk/foi/internalops/hid_circs/technical_general/
spctechgen35.htm

Health and Safety Executive (n.d.e) 'Jet Fires'. Available
at: www.hse.gov.uk/offshore/strategy/jet.htm

Health and Safety Executive (n.d.f) 'Non-Destructive
Testing'. Available at: www.hse.gov.uk/comah/
sragtech/techmeasndt.htm

Health and Safety Executive (n.d.g) 'Pool Fires'.
Available at: www.hse.gov.uk/offshore/strategy/pool.
htm

Health and Safety Executive (1995) *Current Predictive
Methods for Explosion Hazard Assessments
Offshore (Offshore Technology Report)*. London:
HSE Books. ISBN 0-7176-0969-3.

Health and Safety Executive (2000) *Working With
Ionising Radiation*. London: HSE Books. ISBN
978-0-7176-746-3.

Health and Safety Executive (2001) *Decommissioning
Topic Strategy (Offshore Technology Report)*.
London: HSE Books. ISBN 0-7176-2054-9.

Health and Safety Executive (2004) *A Methodology for
the Assignment of Safety Integrity Levels (SILs)
to Safety-Related Control Functions Implemented
by Safety-Related Electrical, Electronic and
Programmable Electronic Control Systems of
Machines*. London: HSE Books. ISBN 0-7176-2832-9.

Health and Safety Executive (2004) 'The Selection and
Use of Flammable Gas Detectors'. HSE. Available at:
www.hse.gov.uk/pubns/gasdetector.pdf

Health and Safety Executive (2009) *Safety and
Environmental Standards for Fuel Storage Sites.
Process Safety Leadership Group Final Report*.
London: HSE Books. ISBN 978-0-176-6386-6.
Available at: www.hse.gov.uk/comah/buncefield/
fuel-storage-sites.pdf

International Marine Contractors Association (2010)
'Guidance on Simultaneous Operations (SIMOPS)'.
Available at: www.imca-int.com/media/73596/
imcam203.pdf

Kletz, T. A. (1998) *What Went Wrong? Case Histories
of Process Plant Disasters*. Oxford: Gulf. ISBN
0-88415-920-5.

Kletz, T. A. (2003) *Still Going Wrong: Case Histories of
Process Plant Disasters and How They Could Have
Been Avoided*. Oxford: Gulf. ISBN 0-75067-709-0.

SINTEF (2007) 'Safety Critical Systems'. Available
at: www.sintef.no/home/Information-and-
Communication-Technology-ICT/Applied-Cybernetics/
Main-fields-of-expertise/Safety-critical-systems

United States Environmental Protection Agency (1997)
'Lightning Hazard to Facilities Handling Flammable
Substances'. Available at: www.epa.gov/osweroe1/
docs/chem/lit-flam.pdf

Varner, V. (2012) 'Pressure Assisted Flares – A User's
Perspective'. Dow Chemical Co. Available at: http://
content.lib.utah.edu/utils/getfile/collection/AFRC/
id/14133/filename/14150.pdf

CHAPTER 17

Element 4

Sub-element 4.1:
Fire and explosion in the oil and gas industries

This chapter covers the following learning outcome

Outline appropriate control measures to minimize the effects of fire and explosion in the oil and gas industries.

17.1 Leak and fire detection systems

17.1.1 Introduction

All process plants, refineries, storage facilities, etc. related to the oil and gas industry should be designed to be as safe as possible. However, no matter how comprehensive the design is, there will always be a residual risk. Consequently, everything possible should be done to mitigate the consequences of any untoward incident, particularly those involving fire or explosion.

The risk of fire and/or explosion on installations carries the highest level of consequence and, as such, needs to be addressed as robustly as possible. This is generally done by designing into the installation a series of protective layers, of which there are usually three:

1 A detection system to warn of any fire or the potential for fire, such as a leak of product, smoke or an increase in localized heat
2 An active layer which aims to deal directly with any outbreak of fire, such as a water deluge system
3 A passive barrier which aims to limit the extent of any fire, whilst the active barrier does its job

We will now look in detail at all three layers.

17.1.2 Detection systems

As we have just mentioned, detecting, controlling and managing leaks, fires and explosions is the first layer of protection from the consequences of fire or explosion. The Fire and Gas (F&G) detection system is a combination of various devices strategically positioned throughout the installation which aim to give the earliest possible warning of any potential problem that could escalate into a fire or explosion if it is not dealt with as quickly as possible.

We are now going to look at the various leak and fire detection systems, including:

▷ **Spot systems**
▷ **Flame detection systems**
▷ **Heat detection systems**
▷ **Gas detection systems**
▷ **Smoke detection systems**
▷ **Leak detection systems**

Detection systems – spot systems

Spot fire detection systems are designed to conduct a continuous thermal surveillance of a specific area. They are sensitive to infrared radiation within a cone of vision.

The extent of the cone of vision, and consequently the area being protected, can vary and is dependent on the model and manufacturer.

Spot fire detection systems are extremely sensitive and are capable of detecting changes in temperature of less than 1°C, although this level of sensitivity can be set to suit individual requirements. This enables the fire and gas system to be forewarned at a very early stage of any potential overheating or fire risk, provided it is within the camera's cone of vision.

Detection systems – flame detection systems

Over and above the heat detection systems we have just covered, there are other systems which are capable of detecting flames. They include:

▷ Camera-based flame detectors
▷ Ultraviolet flame detectors
▷ Combined ultraviolet and infrared flame detectors

We'll now take a brief look at each of these systems to see how they work.

Camera-based flame detectors

Camera-based flame detectors are capable of detecting and pinpointing a fire from long distances as soon as the fire starts. These pictures can be relayed back to the control room operator. A camera-based flame detector can give control room operators real time information about the state of the fire.

These detectors can also make a recording of events prior to the incident. This facilitates a full evaluation of the circumstances surrounding the incident so that proper and appropriate actions can be taken to reduce the risk of a repeat incident.

These types of camera can cover very large areas, which mitigates their cost compared with other fire detection technologies.

Ultraviolet flame detectors

Ultraviolet flame detectors were the original type of flame detector and have mostly been replaced by more advanced technologies. They rely on detecting the ultraviolet radiation produced by flames. However, the more advanced ultraviolet flame detectors are capable of detecting hydrogen fires and, as such, can have a part to play in some fire and gas detection systems.

Combined ultraviolet and infrared flame detectors

These detectors are capable of detecting both ultraviolet radiation and infrared radiation. They tend to be used where the reduction of false alarms is an issue.

Detection systems – heat detection systems

There are two types of heat detector:

▷ Point heat detectors
▷ Linear heat detectors

We'll now take a brief look at each of these systems to see how they work.

Point heat detectors

These systems are capable of detecting a rise in temperature at any point within the range of the detector. There are two types of point heat detectors known as 'rate of rise' detectors and 'fixed temperature' detectors. The rate of rise detector monitors the ambient temperature with a view to detecting any sudden increases in temperature, whilst fixed temperature detectors are set to respond when a preset temperature is reached. Examples of where they might be used include:

▷ Kitchens
▷ Turbine hoods
▷ Pump buildings
▷ Generator hoods
▷ Where use is mandatory through legislation or codes of practice

Linear heat detectors

Linear heat detectors are used where heat detection is required in a linear fashion, for example on tank rim seals or in cable tunnels. They are capable of detecting a fire anywhere along its length.

The linear heat detector comprises a cable which is made up of a pair of twisted wires which are each sheathed in a polymer coating. This coating is engineered to break down at a specific temperature (a full range of temperatures is available to suit varying requirements). If heat or flame melts the polymer coating, this allows the twisted wires to make contact and send a signal to the alarm box and/or control room.

Detection systems – gas detection systems

Gas detection systems are capable of detecting a discharge of combustible or toxic gas. If combustible gas is detected, they will warn the control room of

the incident and may well initiate action from the fire and gas system to control and minimize the leak and prevent it from developing into a fire or explosion.

If a toxic gas is detected, they can warn personnel of the hazard in the area.

There are a number of different types of gas detector available. These include:

▷ Infrared absorption combustible gas detectors
▷ Point infrared gas detectors
▷ Open path infrared gas detectors

We'll now take a brief look at each of these systems to see how they work.

Infrared absorption combustible gas detectors

Infrared absorption combustible gas detectors work by being able to recognize the specific absorption characteristics that hydrocarbon molecules have to infrared light. The more hydrocarbon molecules that are present in a given space being monitored by these devices, the higher the absorption of infrared radiation.

These devices are capable of detecting more than one type of hydrocarbon gas.

Point infrared gas detectors

Point infrared gas detectors monitor a specific location with the aim of measuring the concentration of gas at that location. Consequently, they need to be positioned where a release of gas is considered possible or where gas is considered to be especially problematic, for example the air inlets to temporary refuges.

It is possible to position them remotely and then connect them by tubes to the location requiring monitoring. Air can then be drawn from the sampling location and passed over the detection chamber within the device.

Examples of use are:

▷ Detection in confined spaces
▷ Specific locations
▷ Air inlets into accommodation (e.g. temporary refuge)

Open path infrared gas detectors

Open path infrared gas detectors are made up of two components, a transmitter and a receiver, as shown in Figure 17.1. They constantly monitor the area between these two components to detect the presence of any gas.

They work on the principle of sending two infrared beams between the transmitter and the receiver. The first, a sample beam, is the infrared beam which is set to detect the presence of hydrocarbons. The second infrared beam is set so that its wavelength is outside the gas absorbing range. The two beams are constantly compared and whilst the comparison remains the same as the factory setting, no gas is present and no alarm is raised. However, if the comparison changes because some of the sample beam has been absorbed by gas, the device will trigger an alarm.

These devices can detect more than one type of hydrocarbon gas and different versions can also detect hydrogen sulphide (H_2S).

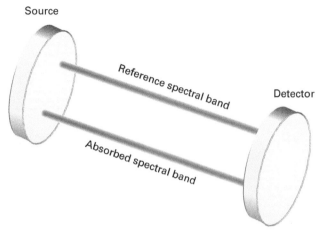

Source

Reference spectral band

Detector

Absorbed spectral band

Figure 17.1 Principle of an infrared beam detector
Source: Wise Global Training.

When a gas cloud crosses the sample beam, the ratio of the two beams changes and this is detected by the system and automatically raises the alarm.

Figure 17.2 shows an open path infrared gas detector units in situ. In this case, the infrared beam has been artificially superimposed in order to highlight its location.

Figure 17.2 Infrared beam detector in situ
Source: Dreamstime.

Detection systems – smoke detection systems

Smoke detectors range from the battery-powered detectors of the type used in the home, to sophisticated camera-based detection systems. They include:

▶ Point smoke detectors
▶ Ionization point smoke detectors
▶ Optical point smoke detectors
▶ Optical beam smoke detectors
▶ Aspiration smoke detectors

We'll now take a brief look at each of these systems to see how they work.

Point smoke detectors

Point smoke detectors are designed to detect smoke at a fixed location. Ideally, they should be positioned where smoke is most likely to accumulate in the event of a fire.

Examples of use are:

▶ Offices
▶ Corridors
▶ Switch rooms
▶ Accommodation

Ionization point smoke detectors

An ionization point smoke detector uses a small radioactive source (a radio isotope) to ionize air inside a detection chamber. Any change in this ionization, due to the presence of smoke particles, will activate the alarm.

In comparison with optical detectors, ionization detectors are more sensitive to the stage of a fire where flames are present, whereas optical detectors are more sensitive to fires when they are still smouldering.

Example of use are:

▶ Where there is the potential for high energy fires
▶ Where use is mandatory through legislation or codes of practice

Optical point smoke detectors

Optical point smoke detectors are designed to detect smoke particles inside a sampling chamber. This is where a source of light is collimated into a beam and a photoelectric sensor is set at an angle to the beam to act as a light detector and monitor it. In the absence of smoke, the beam continues to be emitted in a straight line. However, if smoke particles are present, some of the light is scattered and the light detector senses this and triggers the alarm.

They tend to be used for general purposes but versions which are set to be highly sensitive can give early warnings where this is a requirement.

Optical beam smoke detectors

Optical beam smoke detectors work on the principle of measuring any difference in light emitted from one point and received at another point. Where smoke particles absorb or scatter the light, this is detected and the alarm is triggered.

Optical beam smoke detectors come in two variations – end to end optical beam smoke detectors and reflective optical beam smoke detectors.

End to end optical beam smoke detectors have a separate light transmitter and receiver, whereas reflective optical beam smoke detectors have the light transmitter and receiver within the same unit. The light path is directed from the transmitter to a retro reflector positioned some distance away. The retro reflector then sends the light beam back to the receiver and a comparison is made.

Note: A retro reflector is like a mirror which reflects light to a given location with the minimum of scattering.

Examples of use are:

- Monitoring of large areas, e.g. atria, machinery spaces, etc.
- Areas where it is impractical or impossible to mount other types of detector

Aspiration smoke detectors

Aspiration smoke detectors use a network of tubes to draw samples of air from various locations to a central detection unit. This detection unit then looks for the presence of smoke particles by measuring whether light passed through the samples is scattered or not. If it is scattered, this indicates the presence of smoke and the alarm will be triggered.

Detection systems – leak detection systems

Leak detection systems work on the principle of detecting the ultrasonic sound emitted when a leak of gas or vapour occurs from a small gap, such as those that may develop on valves, flanges and joints in pipework. This sound is inaudible to the human ear but is highly characteristic when gas or vapour leaks from the gap in the plant.

These devices are not limited to any particular gas or vapour, as all gases and vapours emit similar ultrasonic sounds, which makes these devices useful as general gas or vapour leak detectors.

17.1.3 Other design considerations for alarm systems

Now we've looked at the various types of alarm systems available to those who design protective and pre-emptive warning systems into installations, we now need to look at other aspects to be taken into account at the design stage.

Buildings, plant and equipment layout tend to be divided into three categories as far as fire and leak safety engineering is concerned. These are:

- Fire compartments
- Detection zones
- Alarm zones

We will now take a more detailed look at each of these types of zone and see how they affect design decisions.

Design considerations – fire compartments

Buildings are generally divided into sections, known as fire compartments, which use fire resistant structures (walls and floors) to enclose them in order to limit the spread of any fire that may break out within any one of these sections.

It is necessary for the designer of fire detection and alarm systems to be familiar with the design and structure of the building and plant and, in particular, the position and extent of its fire compartments.

Design considerations – detection zones

Detection zones are essentially a convenient way of dividing up a building, area or plant into manageable sections so as to assist in quickly locating the position of any fire or leak.

The zone boundaries do not necessarily have to be physical features of the building, although it is normal practice to make the zone boundary coincide with walls, floors and specifically fire compartments.

The size and position of the detection zones will be more dependent on such things as the extent of areas used for one particular process or the number of people working in one section at a time.

Design considerations – alarm zones

Complex buildings, where it is necessary to operate alarm devices differently in various parts of the building, should be divided into alarm zones so that all of the alarm devices in one alarm zone operate in the same way.

Separate alarm zones are only necessary if different alarm systems are needed in different parts of the same building. If the only requirement is to activate all the alarms in order to provide a single common evacuate signal once a fire is detected, then alarm zones are not needed. The whole building is regarded as one alarm zone.

Alarm zones are not constrained by the extent of detection zones. Alarm zones may contain multiple detection zones.

17.2 Passive fire protection

17.2.1 Introduction

As a general term, passive fire protection covers the materials, products and design measures built into a building or structure in order that any fire which may start in the building or structure is restricted in its growth and spread. This is achieved by controlling the flammability of the structure, including walls, ceilings, floors, doors, etc., as well as by protecting structural steel members from severe heat which might compromise their integrity. Finally, the building or structure is designed in a way that divides it into separate fire containing compartments. This is so that any fire which may start will be restricted, or contained within a limited area.

More specifically, Passive Fire Protection (PFP) can be described as a coating, cladding or free standing system which will act as a thermal barrier to restrict the rate at which heat is transmitted to any object or adjacent area. The type and extent of PFP will depend on a number of factors, including:

- The type of fire anticipated – hydrocarbon jet or pool fires, or cellulosic fires
- The expected maximum duration of fire
- The consequences if the Passive Fire Protection (PFP) is breached

17.2.2 Types of passive fire protection

There are many types of Passive Fire Protection (PFP) materials available. These can be broadly categorized as follows:

- Spray-applied coatings
- Blanket/flexible jacket/wrap around systems
- Prefabricated sections such as walls
- Enclosures and casings
- Composites
- Seals and sealants
- Systems (e.g. cable transit blocks, inspection hatches, pipe penetration systems through bulkheads)

On offshore installations, where weight is at a premium, spray applied coatings are the most frequently used type of passive fire protection.

17.2.3 Aims of passive fire protection

Passive fire protection has three main aims:

1 To prevent steel structures, which are load-bearing, reaching a temperature where their integrity will be compromised. This is generally accepted as being 400°C.
2 To protect process vessels and their supportive structures (legs).
3 To prevent heat transfer through walls, floors and ceilings into adjacent rooms or spaces by limiting the inner wall temperature to no more than 180°C in any one location.

17.2.4 Types of fire

Fires can be categorized as being either cellulosic or hydrocarbon. This refers to the fuel which is being consumed by the fire.

Cellulosic fires are those which burn materials such as timber, upholstery or paper. Hydrocarbon fires are those which burn oils and fuels.

Hydrocarbon fires burn much more fiercely than cellulosic fires. Consequently, Passive Fire Protection (PFP) used in buildings or structures associated with hydrocarbons should be of a higher performance standard than PFP used where the only fire threat is from cellulosic material.

17.2.5 Coding of fire walls

Walls and divisions which have been manufactured as passive fire protection are coded in relation to the type of fire they are designed to withstand and its duration. The code is made up of a letter (either H or A) followed by a number (0, 30, 60 or 120).

The letter indicates what type of fire the wall or division has been made to withstand. 'H' type fire walls are manufactured to a standard that will withstand hydrocarbon fires, whereas 'A' type fire walls are manufactured to a standard that will only withstand cellulosic fires.

The number which follows the letter indicates the length of time the wall or division has been designed to hold back the fire. For example, a wall with a 'H60' coding is designed to withstand a hydrocarbon fire for 60 minutes, whilst a wall with an 'A30' coding is designed to withstand a cellulosic fire for 30 minutes.

17.3 Active fire protection systems – water based extinguishing systems

Active fire protection systems are those fire protection systems which are primed and ready to be activated on a given signal.

17.3.1 Water deluge systems

We have covered water deluge systems in Chapter 13, but in respect of active fire protection systems, it's worth reiterating the issue.

A water deluge system is a means of fighting a fire with large amounts of water. It is generally positioned in areas where hydrocarbons are processed or stored, as well as areas where there is the potential for an uncontrolled release of gas which could result in a fire or explosion.

It works by means of a series of nozzles connected to a piping network. The nozzles are kept permanently in the OPEN position. When an emergency situation arises, the deluge system is activated by the Emergency Shutdown (ESD) system. This switches on the dedicated fire water pumps, resulting in water being emitted from the nozzles and deluging the area.

A deluge system can be activated by the emergency shutdown system or manually.

Figure 17.3 Fire deluge system
Source: Frontline Fire International.

Water deluge systems provide a broad level of protection against a wide range of fire scenarios. This includes offering protection to open or unshielded escape routes.

When activated, deluge systems swamp the target area with water. As well as having a cooling effect on the fire, they also starve the area of oxygen.

Water deluge systems are highly effective when dealing with pool fires, especially when used in conjunction with aqueous film forming foam (AFFF). However, they have a limited effect when required to deal with jet fires, especially those which have high combustion rates.

17.3.2 Sprinkler systems

A fire sprinkler system is an active water based fire protection system consisting of a grid of water pipes, usually located within a roof void, which has sprinkler heads located at regular intervals which protrude through the roof. The sprinklers are spaced in such a way as to give adequate cover to the whole area which needs protecting against the outbreak of fire.

The water in the pipe system remains under pressure (active) but is held back at each sprinkler head by a thermal element. When a thermal element is activated because it has reached its predetermined temperature, it allows water to be sprayed from that sprinkler head. Only sprinkler heads in the vicinity of the fire will be activated, thus ensuring only the immediate area around the fire is sprayed with water. This minimizes any possible water damage.

Sprinkler heads can also be placed in remote areas, such as roof voids and floor ducts where fires may start unnoticed.

The sprinkler head shown in Figure 17.4 has a glass bulb as its thermal element. When the bulb experiences a level of heat equal to its predetermined setting, it will shatter, allowing water to be sprayed.

Figure 17.4 Sprinkler head
Source: iStock.

Sprinkler heads operate individually.

Figure 17.5 Sprinkler head in action
Source: iStock.

17.4 Active fire protection systems – chemical/foam based extinguishing systems

17.4.1 Foam based extinguishing systems

Foam is used as a means to extinguish liquid fires. It has four useful characteristics which make it a suitable medium for fighting liquid fires:

1 It excludes air from the flammable liquids – it takes oxygen out of the fire triangle.
2 It eliminates vapour release from the fuel surface – it takes fuel out of the fire triangle.
3 It separates the flames from the fuel surface – it takes heat out of the fire triangle.
4 It cools the fuel surface.

17.4.2 How foam is made

Foam is made by combining water with a foam concentrate and mixing it in a way that introduces air. It is basically a stable mass of small air-filled bubbles with a density which is lower than oil or gasoline. This means it will readily flow over the surface of burning fuel.

When mixed, foam expands at different rates according to the concentrate used. Low expansion foams, such as Aqueous Film Forming Foam (AFFF) have an expansion ratio of less than 20 times.

Medium expansion foams have an expansion ratio of between 20 and 200 times.

High expansion foams have an expansion ratio of over 200 times. They tend to be used for quickly flooding large enclosed areas.

Figure 17.6 Foam emanating from outlets
Source: Elgin Air Force Base.

17.4.3 **Controlling effects of foam**

Foam can apply a number of fire controlling effects, including:

▶ A separating effect
▶ A covering effect
▶ A cooling effect
▶ A suppression effect
▶ An insulating effect
▶ A film formation effect

Let's look at how each of these effects works.

Controlling effects of foam – a separating effect

The foam cover separates the combustion zone from the surrounding atmosphere and, as such, prevents oxygen feeding the fire.

Controlling effects of foam – a covering effect

The foam cover stops the evaporation of flammable vapours from the burning material. This means flammable vapours, which would normally be encouraged by the heat of the fire and which feed the fire, are stopped at source.

Controlling effects of foam – a cooling effect

When foam is applied to a fire which is fed from a liquid source of fuel, water is gradually discharged by the foam, which adds a cooling effect to the flame.

Controlling effects of foam – a suppression effect

High- or medium-expansion foam used to flood areas will prevent the release of flammable vapours/gases from combining with air, air being necessary for the combustion process to take place.

Controlling effects of foam – an insulating effect

Foam has a low thermal conductivity level. Consequently, when foam covers any flammable liquid which is not burning, it tends to insulate it from thermal radiation and ignition.

Controlling effects of foam – a film formation effect (AFFF: Aqueous Film Forming Foam)

Foam which is used on non-polar hydrocarbons (the majority of hydrocarbons) produces a thin aqueous film which helps the foam flow, as well as assisting in the extinguishing process and inhibiting re-ignition.

17.4.4 **Polymer film additives**

As we've just mentioned, most hydrocarbons are classed as non-polar hydrocarbons and when these burn, the flames can be extinguished by ordinary foam. However, a small group of hydrocarbons, known as polar hydrocarbons, will destroy any ordinary foam used as an extinguishing agent as soon as it is applied.

In order to combat this reaction, an additive called a polymer film former is added to the foam. When the foam with the additive is applied to the burning hydrocarbon, the film floats on top of the hydrocarbon, acting as a barrier between the hydrocarbon and the foam, thus stopping the hydrocarbon from breaking down the foam.

17.4.5 **Chemical fire extinguishing systems**

Chemical fire extinguishing systems use dry powder to extinguish a fire. They can be deployed as hand held extinguishers for use on localized small fires, or as fixed systems to cover a specific area. Dry chemical powder is intended for use in extinguishing fires involving bulk chemical agents and liquefied gases. Chemical fire extinguishing systems are also used to protect areas which are sensitive to water or foam as an extinguishing medium, such as control rooms or areas where electrical or communications equipment is located.

Both the hand held and fixed systems use the principle of introducing an inert gas, usually nitrogen, into the base of the vessel holding the dry chemical. This pressurizes the vessel and allows the dry chemical particles to exit through a discharge valve at a controlled velocity and in a form where they are suspended in the inert gas (a cloud). This release can be directed either manually, or in a pre-arranged direction, onto the fire.

The chemical reacts with the fire and extinguishes it. Monoammonium phosphate based chemicals melt at a relatively low temperature, thus blanketing the burning surfaces and preventing re-ignition.

Chemicals used in dry powder extinguisher systems are non-toxic but can cause skin irritation.

17.4.6 Fixed dry chemical powder fire extinguishing systems on bulk gas carrying ships

In 2009 the International Maritime Organization approved guidelines for installing fixed dry chemical powder fire extinguishing systems on board ships carrying liquefied gases in bulk. The aim is to protect on-deck cargo areas of ships carrying liquefied gases in bulk in accordance with Safety of Life at Sea (SOLAS) regulation II-2/1.6.2 and chapter 11 of the International Code for the Construction and Equipment of Ships Carrying Liquefied Gases in Bulk (IGC Code).

17.5 Active fire protection systems – inert based extinguishing systems

Inert fire extinguishing systems are delivered in two main forms, i.e. inert gas based systems and water mist based systems.

17.5.1 Inert gas systems

Inert gas based fire extinguishing systems are based on the principle that fire needs oxygen in the air to continue its combustion process (burn). The level of oxygen in the air is normally 21 per cent, but if this falls below 15 per cent, the combustion process will stop as there is not enough oxygen present to feed the fire. The use of inert gas aims to reduce the oxygen in the target area to below 15 per cent, thus extinguishing the fire.

The main advantage of inert gas based systems is that they are non-destructive to plant and equipment. This is particularly important where electrical and communication equipment is involved.

The gases used in these systems are carbon dioxide (CO_2), nitrogen (N) and argon (Ar), or a combination of these gases. For example, some systems use a combination of 52 per cent nitrogen, 40 per cent argon and 8 per cent carbon dioxide, whilst another combination is 50 per cent nitrogen and 50 per cent argon.

However, where carbon dioxide is an element of the gas system, it can act as an asphyxiant to anybody in the vicinity, which can prove fatal. Consequently, this has to be a consideration when selecting the best system to use in a particular location.

Nitrogen and argon are less of a problem as they are non-toxic, but their presence will still reduce the oxygen level which, if it falls below 12 per cent, will cause breathing difficulties.

It must be remembered that if and when these systems are activated, no person should be in the area where these gases are being deployed as they may become asphyxiated.

Like chemical fire extinguishing systems, inert gas systems can be deployed as hand held extinguishers, known as a 'local application principle', or fixed systems, known as a 'total flooding principle' for dealing with fires in enclosed spaces or rooms.

Discharge of the gas on fixed systems is through discharge valves positioned at identified critical points within the area being covered by the system.

17.5.2 Water mist systems

Water mist based fire extinguisher systems work on the same principle as a sprinkler system. However, when activated, the water that is discharged is in micron sized droplets which cover a wider area. The droplets rapidly convert the energy in the fire to steam, which starves the fire of oxygen. Because the droplets are so small and abundant, the water absorbs the energy of the fire much faster than a sprinkler system and uses less water. This in turn reduces the potential for any damage that may be caused to the surrounding area.

Water mist also has the effect of partially scrubbing the air of toxic by-products caused by the fire.

Figure 17.7 Mist spray head
Source: Wise Global Training.

17.6 Examples of equipment-specific types of fire protection systems and their functions

17.6.1 Floating roof tanks

Storage tanks with floating roofs have the potential to leak vapour from the rim seal. If that happens and the vapour ignites and it is not dealt with immediately, it can result in an escalation of events and a potential catastrophic outcome.

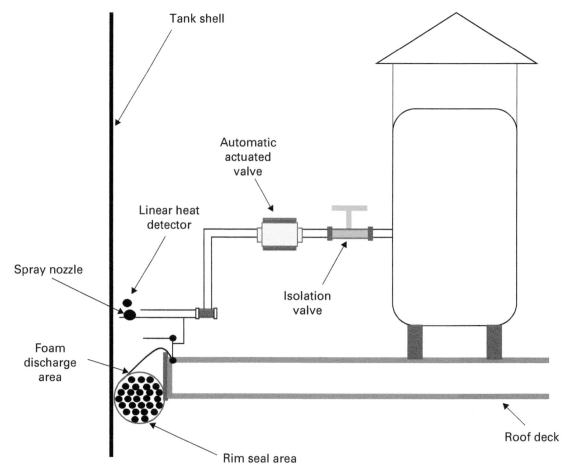

Figure 17.8 Rim seal fire-fighting unit
Source: Wise Global Training.

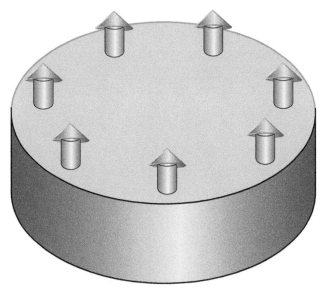

Figure 17.9 Rim seal fire protection overview
Source: Wise Global Training.

Consequently, a constant monitoring and protection system is needed so that any fire is detected and dealt with automatically and immediately. One solution is to place a linear heat detection system at the rim seal area of the roof and connect this to a number of foam based fire extinguishing modules positioned around the perimeter of the roof, as shown in Figures 17.8 and 17.9. When the linear heat detector registers a fire, as

well as raising the alarm, it will also activate the fire extinguisher module nearest to the fire. This will then flood the immediate area with foam and extinguish the fire.

An alternative to individual foam based fire extinguishing modules positioned around the perimeter of the roof is to have a number of foam outlets positioned around the top wall of the tank, as shown in Figure 17.10. These are connected by pipework to a foam mixing tank positioned outside the bunded area. This mixes and pumps the foam through the pipes to the outlets where it is discharged to flood the perimeter of the tank. The foam can be supplemented by the fire service if required as the mixing tank will have a foam inlet for the emergency service to add additional foam.

Additionally, external fixed water spray systems may be fitted. Figure 17.11 shows such a system in operation. As well as fighting fires, this kind of system can offer a cooling effect if required.

17.6.2 Fire protection systems for spherical storage tanks

Fire protection systems for spherical storage tanks generally employ a water based fire protection system. There are a number of systems, each with their own advantages and limitations. These are:

▶ Water deluge systems
▶ Fixed monitor systems
▶ Water spray systems

We shall now look at the merits of each of these systems in turn.

Fire protection systems for spherical storage tanks – water deluge system

A water deluge system for spherical storage tanks works on the principle of having a single or a number of outlets placed on the top of the vessel. When these are activated by an alarm which has sensed a fire, water pours from these outlets and runs down over the surface of the vessel, potentially extinguishing the fire.

The advantages of the system are that it is quick to respond to any alarm trigger, it is generally automatic and, because the outlets are large, it is less prone to plugging.

The disadvantages of the system are that there are areas where the deluge might not prove adequate, such as leg structures, and that this type of system is not as effective against jet fires as other, more direct systems.

Fire protection systems for spherical storage tanks – fixed monitor system

A fixed monitor system for spherical storage tanks is where vessels are surrounded by ground based water hydrants. These either have monitors affixed to them, or monitors are positioned independently around the vessel so that a visual impression is available to the control room at all times.

When a fire is detected, the alarm is raised and the hydrants are either automatically operated by the system or manually operated by personnel. The hydrants can be fitted with an oscillating facility which sprays the water from side to side. This allows the

Figure 17.11 Tank spray system in action
Source: Martin Wall – Shutterstock.

Figure 17.12 Spherical storage tanks
Source: Dreamstime.

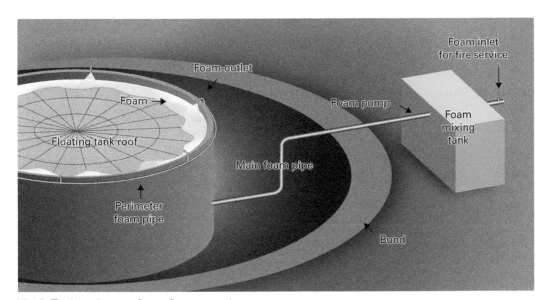

Figure 17.10 Tank perimeter foam fire protection
Source: Tal Glazer.

water to be directed over a wide area without the need for an operator to be present.

The main disadvantage is that if operators have to manually operate the hydrants, this places them dangerously close to the fire. Another factor is that the strength and direction of any wind may have an effect on the water being directed onto the sphere.

Figure 17.14 Spherical tanks with fire protection system
Source: iStock.

Figure 17.13 Fixed point fire-fighting cannon
Source: iStock.

Fire protection systems for spherical storage tanks – water spray system

A water spray system for spherical storage tanks is made up of a network of water pipes which totally surround the vessel. These have spray nozzles fitted to them at regular intervals which point inwards towards the vessel. When these are activated by an alarm which has sensed a fire, water is sprayed from these nozzles onto the entire surface of the vessel with the aim of extinguishing the fire.

The advantages of this system are that it is quick to respond to any alarm trigger, it is generally automatic, and the whole surface of the vessel gets a good dousing.

The disadvantages are that if the vessel experiences a vapour cloud explosion, the system is susceptible to damage and the nozzles, which have small outlets, are

prone to plugging. Also, this type of system is not as effective on jet fires as other, more direct systems.

17.6.3 Gas turbine systems and compressor systems

Gas turbines and other major driven equipment, such as compressors and pumps, are widely used in the oil and gas industry, especially offshore, for a variety of functions. These include generating electricity, air compression, pumping, etc. These items of machinery, particularly gas turbine units, are complex items which operate at high speeds and high temperatures.

A complete gas turbine system is made up of a number of separate areas or compartments. These include the turbine compartment, the generator compartment, the fuel pump compartment, the lubricating skid and the electrical control room, all of which will require individual fire protection systems.

However, the major fire hazard is associated with the turbine compartment, where there is the potential for fuel or lubricating oil to leak and come into contact with surfaces of the turbine, which operate at temperatures well above the auto-ignition temperature of the fuel and oil. If either the fuel or the lubricating oil leaks and comes into contact with one of these hot surfaces, it will cause a fire.

Historically, these types of fire have been dealt with using inert gas systems such as carbon dioxide. However, the effectiveness of these systems is dependent on having an atmospherically secure area so the gas cannot escape, which has proved a problem in the past. One study on gas turbine fires concluded that 37 per cent of fires tackled by inert gas systems failed because the gaseous agent had leaked from protected enclosures through open doors or ventilation.

The present consensus is that water mist systems are the most appropriate means of dealing with a fire in the turbine compartment. However, the main concern lies with ensuring that if the system is activated, it does not cause thermal shock on the turbine casing. This is achieved by strategic positioning of the mist nozzles in a location where they will not directly spray the surface but create a mist cloud which envelopes and cools these hot surfaces by heat transfer, turning the water droplets to steam.

REVISION QUESTIONS FOR ELEMENT 4

1 **Describe** how an infra-red gas detector system works.
2 **Explain** what passive fire protection is.
3 With regard to passive fire protection, **explain** the difference between a wall coded as H120 and a wall coded as A30.
4 Foam has a number of different effects on a fire. **Give THREE** of these effects and **explain** how each of them affects a fire.
5 **Describe** an appropriate fire protection system that could be used on a floating roof storage tank.

You can compare your answers with our guidance at the end of the book under the section entitled 'Revision and assessment guide'.

17

155

CHAPTER 18

Element 4

Sub-element 4.2:
Emergency response

This chapter covers the following learning outcome

Outline the principles, procedures and resources for effective emergency response.

Chapter contents

18.1 Emergency plan

In Chapter 4 we covered safety cases and safety reports and how these provide a convincing and valid argument that a system is adequately safe for a given application. A fundamental element of that safety case/report is having an emergency plan in place.

We are now going to look at the on-site emergency plan which, to be effective, must be based on a full appreciation of the potential for major incidents which will be unique to that particular site or installation.

18.1.1 On-site emergency plan

Operators of both onshore and offshore installations should undertake the following actions as part of the procedure of writing a comprehensive emergency plan:

▶ Identify all the major hazards associated with the operations together with their source, type, scale and consequences. This should include malicious acts.

▶ From these hazards, identify all the conceivable scenarios that could arise which will need an emergency response, including those which involve evacuation, escape and rescue.

▶ From these scenarios, produce a well-defined plan of action which establishes the appropriate response for an emergency situation and which takes into account the varying demands of different scenarios.

▶ Establish the procedures and frequencies required to test and practise the emergency response to be followed in each of the different scenarios identified.

▶ Establish the formal command structure. The person in overall charge in an emergency will be, in the case of an offshore installation, the Offshore Installation Manager (OIM). Onshore it will be a nominated and named competent person until such time as the emergency services take command of the situation.

▶ Establish that those people who will be expected to take an active part in any emergency response, including those in overall command, are competent to do so (competent being defined as having sufficient training and experience and any other relevant qualities).

▶ Establish that there are enough people to respond to any emergency.

▶ Establish the roles and responsibilities of all individuals on the installation.

▶ Establish that there are enough resources to respond to any emergency.

▶ For onshore sites, establish that there are plans for both onsite and offsite should the emergency not be contained within the site.

▶ For onshore sites, establish that a Major Accident Prevention Policy (MAPP) has been prepared and is current.

▶ Establish what measures will be necessary to facilitate a site clean-up and remediation following an incident.

Once all the information listed above has been gathered, the plan itself can be composed.

18.1.2 Contents of an emergency plan

An emergency plan should contain the following information:

Responsibilities and authority of those overseeing an emergency

The command structure for managing the on-site response to an emergency situation in accordance with the planned scheme. This will include:

▶ The names and positions of persons authorized to set emergency procedures in motion.

▶ The name and position of the person in charge of, and co-ordinating, the on-site mitigatory action.

▶ The name and position of the person responsible for liaising with external agencies and/or local authorities.

▶ Details of what arrangements have been made for occasions when senior managers are not available.

▶ The contact details of all authorized personnel.

Types of events planned for and extent of responses planned

This is the principle aspect of an emergency plan. It should include details of the following issues:

▶ The types of emergency situations which have been regarded as reasonably conceivable.

▶ The response strategy for each of these situations.

▶ The details of personnel who have been allocated roles to play in an emergency situation, and their responsibilities.

▶ The details and location of any special equipment, such as fire-fighting equipment and damage control facilities.

Alarm systems and responses to alarms

This should include what alarm systems and arrangements have been made for early detection of a potential emergency situation. This will include what arrangements have been made and responsibilities for ensuring an appropriate response is made by personnel, such as evacuating the area or facility, taking shelter, using protective equipment, etc.

Arrangements for triggering any off site emergency plan

Where an off site plan is applicable, details of what arrangements have been made for alerting off site emergency services and other agencies such as water companies, environmental agency, etc. Also, under what circumstances these alerts should be made, what information will be required by each service or agency, and their contact details.

Training and instructions

This section covers the arrangements made for training staff in their roles and responsibilities in an emergency situation. It also covers the arrangements for, and frequency of, conducting exercises based on all the identified emergency scenarios.

Finally, details of how contractors and visitors will be given instructions of how they should respond in an emergency situation.

Off site communication measures

Establishing who will be responsible for contacting and briefing the media as well as media contact details.

18.1.3 Fire and explosion strategy

Now we have covered the structure of an emergency plan, we should look at how the potential for major events can be reduced or mitigated. This is covered in a document known as a fire and explosion strategy.

A 'fire and explosion strategy' is a combination of measures taken to reduce the risk to personnel in the case of fire or explosion, or that reduces the risk of fire and explosion happening in the first place. Some of those measures listed here will apply specifically to offshore installations, others will apply to onshore installations and the rest to either type of installation.

Measures to be considered in formulating a fire and explosion strategy for a specific installation include:

- Buildings which are occupied should have an assessment made of the risks and hazards they might be vulnerable to if a major incident occurred. From that assessment, appropriate measures should be taken to address those issues. The buildings included in the assessment process should also include temporary and secondary refuges.

- Escape routes should be clearly marked using high visibility signage along their entire route.
- Escape routes should be well lit and include a contingency for emergency lighting in case of power loss.
- All escape routes, where appropriate, should be protected by firewalls or by deluge fire protection systems.
- Escape routes should be of a size that is adequate to accommodate all personnel.
- Where appropriate, the installation should be compartmentalized (have firewalls between compartments).
- Where appropriate, blow out, or explosion panels, should be strategically positioned within the installation to alleviate any overpressure.
- Where appropriate, escape routes should have heat-activated deluge/sprinklers within them.
- Each area of the installation should have more than one escape route.
- Escape routes should be protected against the effects of fire and explosion.
- There should be internal access to the helideck from any temporary refuge facility.
- There should be a policy of ensuring the number of overrides and inhibits applied to the Emergency Shutdown (ESD) system and the Fire and Gas (F&G) system is kept to a minimum.
- At the design stage of an emergency shutdown system, failsafe and fireproof ball valves should be incorporated so their integrity will not be compromised in the case of a fire.
- At the design stage of a process system, the amount of flanged pipework should be minimized in order to reduce the potential for leaks.
- At the design stage the inventory of hazardous substances should be reduced to a minimum.
- Emergency Shutdown Valves (ESDVs) should be enclosed with fireproof casing.
- Water deluge operating skids should be situated away from the area they are protecting.
- Access doors to accommodation areas should have automatic door closers fitted to prevent ingress of smoke and flames.
- All enclosures which house rotating equipment and electric drives should have Very Early Smoke Detection Apparatus (VESDA) fitted.
- Measures should be taken to ensure the mechanical and natural ventilation to production areas is sufficient to assist in dispersing any gas leak.
- The accommodation and control rooms should be segregated and distanced away from production processes wherever possible.
- The control rooms and emergency command and control centres should be segregated using blast and fire walls.
- Subsea Isolation Valves (SSIVs) should be fitted in sea lines and wells.

18

- ▶ High Integrity Pipeline Protection Systems (HIPPS) should be fitted where appropriate.
- ▶ External fire protection should be fitted to the accommodation rooms and Temporary Refuge (TR).
- ▶ The temporary refuge should be airtight and always under positive atmospheric pressure.
- ▶ There should be a separate Emergency Command and Control (ECC) centre in the temporary refuge when the control room is not situated within the TR.

18.2 Alarms – importance of response

18.2.1 Onshore alarms

When an emergency situation arises within an onshore facility, it is imperative that everybody on the site, as well as the general public in the vicinity, is made aware of the situation. This alarm signal is conveyed by means of a warning siren which is loud enough for everybody on the site, and in the surrounding area, to hear.

It is the responsibility of the site management to ensure everybody knows, and can recognize, the warning siren when it is sounded and what it signifies. This includes the general public who may live or work in the surrounding area. Consequently, well-publicized tests of the siren should be conducted periodically to raise awareness. Finally, so there can be no misunderstanding of why the siren is being sounded, the siren will be of one type only and for all incidents.

There will be areas within the facility that are particularly noisy, such as machinery spaces and enclosed rooms. In these instances, the alarm should be backed up with flashing beacons.

The alarm is also likely to be linked to the control centre of the emergency services so that they are automatically made aware of any emergency.

Training

All personnel will need to undergo training in how to respond when an alarm is activated. This will include practising evacuation drills and how to leave the site in a controlled manner, including giving the correct notification when leaving the site. This aspect is essential for those who are managing the situation as it allows them to know who can and cannot be accounted for.

18.2.2 Offshore alarms

Offshore, the alarm situation is somewhat different as there are two types of alarm used. The first type of alarm is the General Platform Alarm (GPA), which is a general alarm calling all Personnel On Board (POB) to go to their allocated muster station. The General Platform Alarm (GPA) is an intermittent signal of a constant frequency.

These alarms are now standardized throughout platforms operating within the United Kingdom Offshore Operators Association (UKOOA) in accordance with the Prevention of Fire, Explosion and Emergency Response Regulations (PFEER), but may differ internationally.

The second type of alarm is the Prepare to Abandon Platform Alarm (PAPA), which is sounded to inform personnel of the imminent evacuation of the platform. The PAPA is a continuous signal of variable frequency.

Both types of alarm are generally backed up with a public address announcement and by a visual alarm system where necessary. Furthermore, communications will be established between persons who are engaged with activities associated with the installation such as diving operations, supply vessels, aircraft, Mobile Offshore Drilling Units (MODUs) and Standby Vessels (SBVs).

18.3 Medical emergency planning, tiered response, medical evacuation procedures and back up resources

Medical emergencies that occur within oil and gas installations are dealt with differently depending upon whether they are onshore or offshore.

18.3.1 Medical response – onshore

Onshore, where a medical emergency occurs, the control and command team will be able to call upon the local ambulance service to deal with the casualties. Once they arrive on scene they will take control of managing the casualties by triaging them in order to prioritize their medical needs. Once the immediate medical needs have been dealt with by the paramedics, they will be conveyed to hospital as required for more specific medical attention. Prior to the ambulance service arriving, casualties will be looked after by the first-aid team.

Onshore emergency teams, which include everybody involved in contributing to manage emergency medical situations, should regularly be involved in practices with outside agencies so they can handle multiple casualty situations efficiently and effectively.

18.3.2 Medical response – offshore

Offshore, the situation is different because the support from external emergency services has much greater time constraints associated with it. Consequently, in the first instance, casualties will be dealt with by the platform medic, who will be trained in conducting triage of casualties. Triage is a means of prioritizing casualties in relation to their medical needs so the most severe casualties can be dealt with first. The platform medic will also have the use of a fully equipped medical suite and a number of first-aiders to help him/her.

Depending on the location of the installation and pre-arranged arrangements for dealing with medical emergencies, the platform medic may also be able to call upon the service of an onshore doctor for advice. Some areas also have dedicated medical centres to call on for assistance. For example, in the North Sea, the North Sea Medical Centre (NSMC) can be contacted for assistance by telephone or video link at any time.

As well as the platform medic, there is usually a person on the platform's standby vessel with Advanced Medical Aid (AMA) training who can be called upon if needed.

Search and Rescue (SAR) helicopters can also be called upon to medivac (medical evacuation) casualties off the installation. Search and rescue helicopters carry a paramedic on board who will attend to the immediate

medical needs of the casualties. They will also assist in triaging the casualties where needed.

Once the immediate medical needs have been dealt with by the paramedics, those casualties needing further medical attention will be conveyed to hospital by the search and rescue helicopter.

In any offshore medical emergency situation, the Emergency Command and Control (ECC) centre will always contact the coastguard and scramble a search and rescue helicopter to help in managing the casualties.

18.3.3 Triage

We've just mentioned that triage is a process of prioritizing casualties according to their medical needs. Let's now have a look at what this process is in more detail in Figure 18.1.

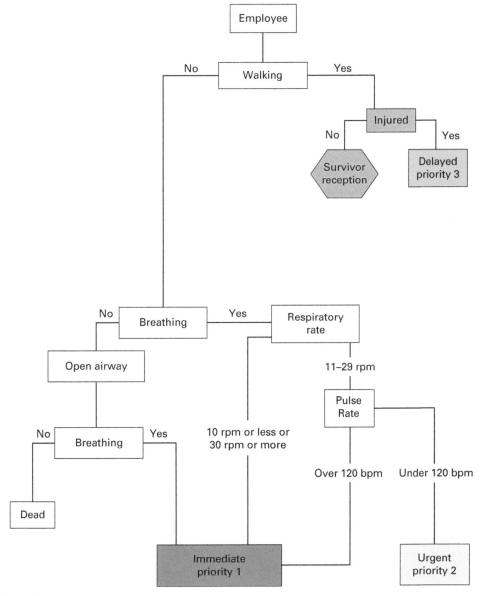

Figure 18.1 Triage flow chart
Source: Wise Global Training.

Telephone triage

Offshore, telephone triage is another way casualties can be assessed and prioritized in the order of severity of their conditions. This takes the form of a set of questions asked over the telephone by onshore medical staff about the extent of injury to each casualty. This allows a checklist to be completed, with each answer being given an appropriate score. The scores then allow the onshore triage staff to prioritize and place each casualty in their appropriate category of treatment.

18.4 Principles of escape, evacuation and rescue from onshore facilities and offshore platforms

The process of evacuating an oil and gas installation varies greatly between those situated onshore and those situated offshore. Consequently, we need to look at the evacuation process of both types of installations separately.

18.4.1 Escape and evacuation – onshore

When personnel need to escape or evacuate an onshore installation there are a number of factors which can enhance their ability to escape without undue difficulty. These include:

▶ Escape routes should be clearly marked using high visibility signage along their entire route.
▶ Escape routes should be well lit and include emergency lighting in case power is lost.
▶ All escape routes should be protected, where possible, by firewalls or by deluge fire protection systems.
▶ Escape routes should be congestion free (have clear access and egress) and be adequate in size to accommodate all personnel.
▶ Escape routes should have heat activated deluge/ sprinklers within them.
▶ Each area of the installation should have more than one escape route.
▶ Where appropriate, the installation should be compartmentalized (have firewalls between compartments).
▶ Where appropriate, blow out, or explosion panels, should be strategically positioned within the installation to alleviate any overpressure.

When an incident occurs within an onshore installation, the normal means for personnel to evacuate the site would be through the main exits. However, there may be instances where these exits have become blocked or are unavailable. Where this is the case, escape and evacuation will have to be by an alternative exit and this will have to have been foreseen and planned for as part

of the planning process when the original emergency plan was drawn up and different scenarios were being considered.

The emergency plan will set out procedures for all conceivable eventualities, and the Emergency Command and Control (ECC) centre will follow these procedures in conjunction with the emergency services.

Evacuation of casualties will be by the most appropriate means available, usually by ambulance. In certain circumstances, an air ambulance may be deemed more appropriate, especially where time is of the essence with seriously injured persons. However, other factors such as risk of explosion or toxic fumes may influence the decision to use helicopters. Visibility (night-time or foggy conditions) or weather conditions (strong winds or thunderstorms) will also affect this decision.

The local authority should work with the operators of the installation in preparing an off-site emergency plan. This will set out how they will respond to an incident which affects the surrounding area, its population and the environment. It will include how the area will be evacuated if needs be, and their response to, and rescue from, damaged property.

18.4.2 Escape and evacuation – offshore

When personnel need to escape or evacuate an offshore installation, there are a number of factors which can enhance their ability to escape without undue difficulty. These include:

▶ Escape routes should be clearly marked using high visibility signage along their entire route.
▶ Escape routes should be well lit and include emergency lighting in case power is cut off.
▶ All escape routes should be protected, where possible, by firewalls or by deluge fire protection systems.
▶ Escape routes should be congestion free (have clear access and egress) and be adequate in size to accommodate all personnel.
▶ The installation should be compartmentalized (have fire walls between compartments).
▶ Blow out, or explosion panels, are strategically positioned within the installation to alleviate any overpressure.
▶ Escape routes should have heat activated deluge/ sprinklers within them.
▶ Each area of the installation should have more than one escape route.
▶ Temporary Refuge buildings (TR) should be constructed from material which has a fire rating of at least H120. This will give a 2-hour protection period before having to evacuate or abandon the installation.
▶ Temporary refuge buildings should be under positive atmospheric pressure using an airlock system.

- There should be more than one means of communicating to personnel specific instructions, such as what to do and where to go.
- The number of ways an alarm is conveyed to personnel should not be by siren alone; i.e. a flashing beacon should be used for areas where a siren might not be heard.
- Multiple means of manually descending to sea level are provided, i.e. knotted rope, sea ladder attached to the platform leg, scramble net, skyscape, etc.
- Appropriate personnel escape equipment is available either in the accommodation area or on each escape route. This equipment should include Emergency Breathing Systems (EBS), Emergency Life Support Apparatus (ELSA), smoke hoods, torches and flame retardant gloves.

When an incident occurs on an offshore installation, and escape and evacuation are required, there are a number of means of leaving the installation safely.

Figure 18.3 Life rafts
Source: iStock.

Life rafts are another means of leaving the installation safely. However, they are not as efficient or as quick to escape in as lifeboats and should rank as second in choice as a means of escape.

Life rafts have to be launched into the sea by means of a davit before they are inflated and can be boarded, which means personnel have to get down from the platform to sea level in order to board the life raft. This can be by 'donut', which is a personnel-controlled descent device. This device is attached to a rope which is attached to the platform. The person controls his/her descent with the device and another rope, attached to the life raft, guides the person to the life raft.

Other means of descending to sea level in an emergency include by knotted rope, by sea ladder attached to the platform leg, by scramble net, or by 'skyscape' (a ladder type escape device).

Figure 18.2 Free-fall lifeboat
Source: Dreamstime.

The primary method is by lifeboat. These can be launched by davit (a crane-type device which lowers the lifeboat to the sea) or by free-fall where the lifeboat is set at an angle on a launch ramp and allowed to fall into the sea when required.

Figure 18.4 Helicopter approaching oil rig
Source: iStock.

18

163

Escape can also be considered by helicopter. However, this is usually restricted to casualties who are not capable of leaving the installation unaided.

There is also the consideration of the extent of the incident, which might restrict helicopters landing on the platform.

Figure 18.5 Standby vessel
Source: Dreamstime.

The platform's standby vessel will be available to respond to an escape and evacuation situation. As well as assisting with the evacuation of the platform, it can offer sea rescue with its Fast Rescue Craft (FRC) and receive launched survival craft (lifeboats and life rafts). If needed, it can also use its fire-fighting capabilities.

It is the duty holder's responsibility to ensure appropriate and comprehensive arrangements are in place for the effective recovery of persons involved in escape or evacuation from the installation. This includes persons falling into the sea from the installation (man overboard), or a helicopter ditching close to the installation.

On recovery, they must be taken to a place of safety. This can be a location onshore, or offshore, for example the platform's standby vessel, where medical treatment and other care facilities can be made available.

18.5 Roles and operation of fire teams onshore and offshore in upstream and downstream facilities

18.5.1 Onshore fire teams

Historically, many onshore installations have had their own in-house fire and rescue teams and equipment, including fire tenders, so that any emergency situation could be responded to quickly. The fire teams are made up of on-site personnel who would normally be doing other jobs around the site, but additionally are trained in fire-fighting and search and rescue techniques.

When called upon to deal with an incident, the role of the on-site fire team is to control the situation as far as possible and make an evaluation to pass onto the emergency services when they arrive and take over command.

In recent years, the role of controlling and containing fires has moved more towards automatic, remote and fixed fire fighting systems such as water deluge systems, sprinkler and mist systems, foam based systems, chemical based systems and inert gas systems. This has been bolstered by the application of passive fire defence systems such as incorporating fire walls into structures and the sectioning and compartmentalization of areas. These measures are expected to contain a situation until emergency services arrive on site and take command.

Consequently, the need for site personnel to become directly involved in fighting large fires has been generally superseded. Response teams are only expected to conduct search and rescue operations should personnel be unaccounted for, and are trained in the use of Breathing Apparatus (BA).

18.5.2 Offshore fire teams

Offshore, the industry has also moved away from personnel directly fighting hydrocarbon fires, to one where on-board automatic, remote and fixed fire-fighting systems are expected to control, contain and bring to a satisfactory end almost any emergency situation. However, the platform's standby vessel may be able to assist with water cannon if this is deemed appropriate.

However, if the situation cannot be contained and brought to an end satisfactorily, the next level of response is to evacuate and abandon the installation.

As with their onshore counterparts, offshore response teams now have the role of search and rescue for unaccounted-for personnel, and are trained in the use of Breathing Apparatus (BA).

The offshore installation has a limited amount of external assistance it can call upon in an emergency, most of which have a significant time lag in how quickly they can respond. These services include:

► The platform Standby Vessel (SBV)
► Coastguard and search and rescue services (helicopters and lifeboat services)
► Commercial helicopters
► Supply vessels
► Shipping in the vicinity

18.6 Training and drills

Training and drills, in relation to emergency response, are about ensuring everybody is in a state of preparedness and knows exactly what to do and what will be expected of them should an emergency situation arise. We will now take a look at training and drills separately.

18.6.1 Training

The oil and gas industry was greatly affected by the Piper Alpha disaster of 1988, which killed 167 people. Tragic as it was, many good things came about as a result of the inquiry and report which followed. The report made 106 separate recommendations, a significant number of which have had a direct impact on emergency response and rescue training since then.

The Prevention of Fire, Explosion and Emergency Response (PFEER) regulations came about as a direct result of the report forthcoming from the Piper Alpha disaster, the principles of which have been generally accepted on a global basis by the oil and gas industry. The regulations relate to both onshore and offshore installations.

The regulations which are pertinent to emergency situations state the following:

▶ People in command should be competent.
▶ There should be sufficient number of competent people on the installation to undertake emergency duties and operate relevant equipment, including sufficient numbers of people to be in attendance at the helicopter landing area during helicopter movements.
▶ Everybody on board an offshore installation should have undergone general training in emergencies, including training in personal survival, installation-specific induction training and training based on the emergency response plan.

Competence can be defined as 'having sufficient training, knowledge and experience to undertake a task or duty safely and effectively'.

18.6.2 Training specific to offshore installations

For all personnel on offshore installations there are specialist training requirements for which standardized training courses have been developed.

Everyone who is working offshore has to undergo general training in emergency situations. This is known officially as the Basic Offshore Safety Induction and Emergency Training (BOSIET). This basic training covers:

▶ First aid
▶ Basic fire-fighting
▶ Self-rescue
▶ Helicopter safety and escape
▶ Sea survival

Beyond the basic course, which everybody has to undertake, more specific courses have been developed.

for designated personnel who will play a specific part in any emergency situation. One of these courses is for coxswains who will be in charge of lifeboats. This course is known as Totally Enclosed Propelled Survival Craft course (TEMPSC). The training these coxswains receive should be relevant to, and match, the type of craft they will be in charge of.

Other specific courses exist for

▶ Muster co-ordinators
▶ Muster checkers
▶ Radio operators

For personnel who are crew members of the emergency response and rescue vessel, a suite of standardized competency-based training has been developed covering:

▶ Initial shipboard operations
▶ Fast rescue craft boatman
▶ Fast rescue craft coxswain
▶ Daughter craft coxswain
▶ Advanced medical aid

18.6.3 Drills

Drills, in relation to emergency situations, are exercises which are undertaken to evaluate emergency plans and procedures under realistic conditions. They also serve the purpose of training personnel in the practical application of their roles and responsibilities. Consequently, they should be conducted as often as is deemed necessary in order to maintain competency levels.

Involving external emergency services and agencies in these exercises is considered essential in order to facilitate an efficient and effective response should the need arise.

Where possible, these exercises should be overseen by an external organization which can independently evaluate performances under near-realistic conditions. This independent evaluation will add credibility to any level of competency attributed to exercise evaluations.

Offshore, muster drills should be undertaken weekly. However, these drills should be random, unannounced and undertaken at varying times of the shift in order to prove they are effective under all foreseeable circumstances.

18.7 External support agencies and resource liaison, including municipal and offshore

18.7.1 Strategic command posts

Any major emergency will involve the external emergency services. If it's an onshore installation this will be fire, ambulance and police. If it's an offshore installation this will be the coastguard agency and the police. The coastguard agency will involve other support agencies such as air–sea rescue and lifeboat rescue as required.

As we mentioned earlier, these services will be regularly involved in drills and exercises with the site's response team, which allows them to practise dealing with all identified major hazards.

When a major incident occurs, the emergency services will establish a set of strategic command posts to oversee and manage the situation. These are likely to be made up of:

▶ The main HQ commanders responsible for determining the best strategy for dealing with the incident. These commanders are generally not located at the incident and, if various organizations are involved, may well be in two or three locations. They will, however, be in constant touch with each other by video conference or by telephone.

▶ Strategic commanders are the next level down, possibly located at the rally point for the emergency services. These commanders are responsible for converting the strategy set out by the HQ-based main commanders into a set of actions or tactics, to be implemented by the emergency teams at the scene.

▶ The final level of command is the supervisors in charge of the emergency workers at the scene. They will be in touch with the strategic commanders to find out what strategy and/or tactics have been established as the best way to deal with the incident. They will also feed back information on how well, or otherwise, these strategies/tactics are working.

18.7.2 External support agencies

Other external assistance might include:

▶ Local councils
▶ The health and safety authority
▶ The agency responsible for the environment
▶ water/drainage company
▶ Coastguard
▶ Air–sea rescue
▶ Lifeboat rescue

18.8 Liaison with emergency services

In order to ensure any emergency situation is responded to effectively, it is essential that good channels of communication are established between the management of a facility and the emergency services. The groundwork for these channels will be set as part of the drills and exercises the parties jointly take part in, but over and above these activities, a

constant liaison with the emergency services should be established and maintained. This should be done through formal channels on a structured basis to emphasize the importance it commands.

When an emergency situation is responded to by the emergency services, there is a certain amount of information which will be required immediately. It is the responsibility of the person who is appointed as the emergency services liaison officer to ensure this information (a) has been provided prior to the incident as part of normal liaison activities, or (b) is ready to be provided as soon as the emergency services arrive on site as part of the update of the situation. That information will include:

1 Contact point for the fire/police/ambulance liaison officer
2 Contact point for fire/police/ambulance incident commander
3 Rendezvous point for emergency services
4 Strategic response group member (planning co-ordinator or as otherwise delegated) with primary responsibility for managing the emergency at the site
5 Likely cause and effect of the emergency
6 Likely casualty status including potentials, how serious and their current location
7 Roll call results
8 Map of the site including floor plans and entry and exit points
9 Evacuation location
10 Outline of the local environment and surrounding risks (possibility of secondary incidents/contamination)
11 Utility shut-off points
12 Availability of CCTV
13 Press liaison details
14 Welfare arrangements
15 Traffic control points and likely impacts on the surrounding area

---

REVISION QUESTIONS FOR ELEMENT 4 CONTINUED

6 **Give FOUR** factors which can enhance chances of survival and escape from an installation in an emergency situation.
7 **Explain** what emergency drills are and who should participate in them.

You can compare your answers with our guidance at the end of the book under the section entitled 'Revision and assessment guide'.

Sources of reference for Element 4

American Petroleum Institute (2001) *API RP 2021 Management of Atmospheric Storage Tank Fires 4th edition*. Washington, DC: API.

British Standards Institution (1999) *ISO 13702:1999 Petroleum and Natural Gas Industries – Control and Mitigation of Fires and Explosions on Offshore Production Installations – Requirements and Guidelines*. London: BSI. ISBN 0-580-32342-0.

Buncefield Major Incident Investigation Board (2008) *Final Report Volume 2: Recommendations on the Emergency Preparedness for, Response to and Recovery From Incidents*. London: HSE Books. ISBN 978-0-7176-6318-7.

Health and Safety Executive (n.d.a) 'Active/Passive Fire Protection'. HSE Guidance. Available at: www.hse.gov.uk/comah/sragtech/techmeasfire.htm

Health and Safety Executive (n.d.b) 'Fire and Gas Detection'. Available at: www.hse.gov.uk/offshore/strategy/fgdetect.htm

Health and Safety Executive (n.d.c) 'Methods of Approximation and Determination of Human Vulnerability for Offshore Major Accident Hazard Assessment'. Available at: www.hse.gov.uk/foi/internalops/hid_circs/technical_osd/spc_tech_osd_30/spctecosd30.pdf

Health and Safety Executive (1997) *Prevention of Fire and Explosion and Emergency Response on Offshore Installations L65*. 2nd edn. London: HSE Books. ISBN 978-0-7176-1386-1.

Health and Safety Executive (1999) *Emergency Planning for Major Accidents*. London: HSE Books. ISBN 978-0-7176-1695-4.

Health and Safety Executive (2003) 'Management of Collision Risk: Radio Communication Between Offshore Installations, Their Standby Vessels and Merchant Ships Operations Notice 61'. HSE Guidance. Available at: www.hse.gov.uk/offshore/notices/on_61.htm

Health and Safety Executive (2007) 'Advice on Acceptance Criteria for Damaged Passive Fire Protection (PFP) Coatings'. Offshore Information Sheet No. 12/2007. Available at: www.hse.gov.uk/offshore/infosheets/is12-2007.pdf

Health and Safety Executive (2009) *Safety and Environmental Standards for Fuel Storage Sites The Process Safety Leadership Group Final Report*. London: HSE. ISBN 978-0-7176-6386-6. Available at: www.hse.gov.uk/comah/buncefield/fuel-storage-sites.pdf

Health and Safety Executive Hazardous Installations Directorate – Offshore Division (n.d.) 'Fire and Explosion Strategy Issue 1'. Available at: www.hse.gov.uk/offshore/strategy/fireexplosion.pdf

Hind, J. (2009) 'Fire and Gas Detection in the Process Industry'. Available at: www.jonhind.com/fire-and-gas-in-the-process-industry-jon-hind%20paper.pdf

International Association of Oil and Gas Producers (2000) *Fire Systems Integrity Assurance: OGP Report No. 6.85/304*. Available at: www.ogp.org.uk/pubs/304.pdf

International Association of Oil and Gas Producers/IPIECA (2011) *Managing Health for Field Operations in Oil and Gas Activities: OGP Report Number 343*. Available at: www.internationalsosfoundation.org/?wpfb_dl=81

International Maritime Organization (2009) 'Guidelines for the Approval of Fixed Dry Chemical Powder Fire-Extinguishing Systems for the Protection of Ships Carrying Liquefied Gases in Bulk'. Available at: www.imo.org/blast/blastDataHelper.asp?data_id=25958&filename=1315.pdf

Roberts, T. A., Medonos, S. and Shirvill, L. C. (2000) *Offshore Technology Report OTO 200 051: Review of Response of Pressurised Process Vessels and Equipment to Fire Attack*. London: HSE. Available at: www.hse.gov.uk/research/otopdf/2000/oto00051.pdf

Wall, M., Lee, R. and Frost, S. (2006) *Offshore Gas Turbines (and Major Driven Equipment) Integrity and Inspection Guidance Notes (RR430)*. London: HSE Books. Available at: www.hse.gov.uk/research/rrpdf/rr430.pdf

18

CHAPTER 19

Element 5

Sub-element 5.1:
Marine transport

This chapter covers the following learning outcome

Identify the main hazards of, and suitable controls for, marine transport in the oil and gas industries.

19.1 Hazards of vessels and working over water

19.1.1 Introduction

The part that shipping plays in the transportation of hydrocarbons around the world, on behalf of the oil and gas industry, is substantial. Shipping also plays an essential part in providing necessary support and services to offshore installations. Consequently, marine activities present many unique risks and hazards which require special consideration in order to control them.

Apart from those risks and hazards associated with dealing with hydrocarbons in a marine environment, unassociated activities by vessels not connected with the oil and gas industry also present a serious risk to offshore installations. These include potential collisions with installations or support vessels, or anchors being dragged over well heads or subsea pipelines, to name just two.

Also, many offshore installations are located in or near busy shipping lanes, which exacerbates the problem of ships straying into the exclusion zone surrounding each of these installations.

As dangerous as passing vessels are, the majority of collisions with offshore installations involve attendant vessels. Attendant vessels cause around ten times more severe damage collisions than passing vessels and this can result in catastrophic losses. Let's take a look at case study 19.1, involving an attendant vessel and an offshore installation.

In order to minimize the risks associated with attendant vessels colliding with installations, support vessels should generally work on the lee side (downwind) of the installation. If support vessels have to work with the rig on a side other than the lee side, a full risk assessment should always be conducted prior to the work commencing.

19.1.2 Exclusion zone

The immediate area around an installation is regarded as a major risk area, with potentially severe consequences for any incident that happens within it. Consequently, it is designated as an exclusion zone with restrictions on which vessels are allowed to enter the zone.

The zone extends for 500 metres around the installation and is constantly radar monitored, as well as being patrolled by the platform's Standby Vessel (SBV) or its emergency response and rescue vessel, either of which is in close communication with the platform's Central Control Room (CCR). Any vessel wishing to enter the exclusion zone must seek permission from the central control room and the patrol vessel before doing so. If a vessel tries to enter the zone without permission, it will be warned off by the patrol vessel.

There are several types of vessel which are usually permitted to enter the exclusion zone. They are:

► Supply vessels
► Mobile Offshore Drilling Units (MODUs)
► Dive Support Vessels (DSVs)
► Coastguard vessels
► Trinity House survey vessels

19.1.3 Vessel hazards

Let's now look at some of the other hazards associated with vessels. These include:

► Breakdown, loss of power or loss of steering. This can lead to drifting, collision, running aground, etc.
► Anchoring over pipelines, wells and submerged cables. This can lead to damage or rupture of pipelines, wells or cables.
► Explosion during loading/unloading operations.
► Pollution – spillage, leakage, etc.
► Striking the installation (e.g. by a platform supply vessel in adverse weather).
► Man Overboard (MOB). The personal hazards associated with someone who falls into the water include:
  ▷ drowning
  ▷ hypothermia
  ▷ being struck by debris or vessel
  ▷ becoming entrapped by debris.

19.1.4 Types of vessels and activities associated with the oil and gas industry

We are now going to look at the different activities associated with the oil and gas industry and the different vessels used to undertake those activities.

Types of vessels – platform support vessel

The Platform Support Vessel (PSV) acts as a shuttle between the offshore platform and the mainland,

---

Case study 19.1

Mumbai High North Disaster 2005 – In 2005 the support vessel, MSV *Samudra Suraksha*, was in the process of transferring an injured worker to the Mumbai High North rig for medical attention when strong sea swells caused the vessel's helideck to strike and sever one of the gas export risers. The escaping gas soon ignited and quickly developed into a fire which engulfed the whole installation.

Twenty-two crew members were lost in the incident and the rig was completely destroyed by fire after two hours, with only the stumps of the jacket being left visible.

bringing all the goods, materials and spare parts required to keep the platform operational. It also takes back to the mainland any waste material and equipment in need of repair.

The PSV has to work in close proximity to the platform in order to unload and load its cargo. Hazards include:

▶ Collision between the platform support vessel and platform.
▶ Lifting hazards from crane operations. This includes shock loading, which is where craning operations are overtaken by the swell of the sea lifting the PSV.
▶ Dropped objects from the platform onto the PSV.

Figure 19.1 Platform support vessel
Source: Dreamstime.

Types of vessels – floating assets

If the results from any test well drilling operations show that there is not enough hydrocarbon reserve to justify building a fixed production platform, then positioning a Floating Production, Storage and Offloading (FPSO) unit over the well head may be an alternative option.

An FPSO unit is a vessel used for processing and storing oil and gas extracted from the well head. This is then offloaded at regular intervals to tankers for transportation to onshore terminals/refineries.

FPSOs can be converted oil tankers, or can be vessels specially built for the purpose. In recent years, purpose built FPSOs specifically designed to store Liquefied Natural Gas (LNG) have come into use. They have a liquefaction facility on board which converts the gas (methane) into a liquid. It does this by reducing the temperature of the gas to −165°C. In its liquid form, methane takes up 1/600th of the space it does when it is in a gaseous form.

Figure 19.2 shows an FPSO on station with its production and service connections.

Offloading operations

The transfer of hydrocarbon product from the floating production, storage and offloading unit to the tanker is a hazardous operation, with the main risk being that

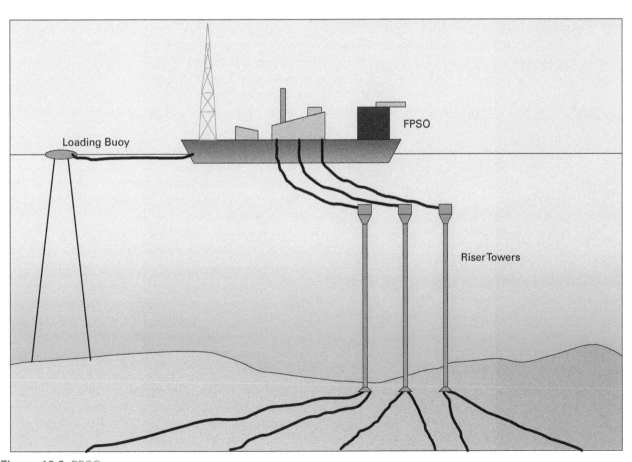

Figure 19.2 FPSO
Source: Wise Global Training.

171

the two vessels may come together as the tanker manoeuvres into position. In order to reduce this risk of collision, Yokohama fenders, like the one shown in Figure 19.3, are placed between the two vessels so they do not make contact with each other.

Figure 19.3 Yokohama fender
Source: Blue Ocean Tackle Inc.

19.1.5 Single buoy mooring

There are many remotely situated sea-based well heads throughout the world which only have a buoy connected to them. These are known as Single Buoy Moorings (SBM) or Single Point Moorings (SPM). These buoys (Figure 19.4) act as a mooring point for tankers and have within them a product transfer system which facilitates the transfer of hydrocarbon product to the tanker.

Figure 19.4 Single buoy mooring
Source: Leighton Offshore.

The main hazard associated with this operation is the potential for the tanker and buoy to come together. The main control for this hazard is to have a support vessel secured to the stern of the tanker to hold the tanker off the buoy.

19.2 Loading and unloading of vessels at marine terminals

19.2.1 Transfer of material between marine vessels and tanks

When a vessel arrives at a port for loading operations it will usually be navigated to its berth by a port pilot who will have the local knowledge and experience of the harbour waters to make this final leg of the journey safely. The manoeuvre may require the assistance of tug boats but many ships now have bow thrusters to make the final positioning adjustments themselves.

Once the vessel is safely moored it will be rigged with 'fire wires'. These are two wire ropes, one fitted fore and one aft on the outboard side of the vessel. This is to provide an immediate hook up for a tug if it needs to tow the vessel away from its berth in the case of an emergency, such as a fire.

Before loading operations commence, a loading plan is formulated which specifies:

▶ How much cargo is to be loaded
▶ The de-ballasting plan
▶ The names of key personnel involved in loading operations
▶ Emergency shutdown, spill procedures and watch arrangements

Thereafter, fire-fighting equipment is put in place; openings on a ship's deck which allow water to drain away, known as scuppers, are plugged to prevent any accidental spillage of cargo running over the side; an emergency shutdown device is positioned at a convenient location, usually at the manifold, so that an emergency shutdown can be activated quickly should something go wrong.

Loading will commence once agreed by the terminal and the vessel. The maximum loading rate and back pressure will be set and good communications will be established and maintained between the terminal and the vessel at all times.

Loading starts at a reduced rate to reduce the generation of static electricity. The ullage in the first tank to be filled is monitored to ensure filling rates conform to the loading plan. Close watch is kept on the manifold back pressure until the agreed maximum loading rate agreed is reached. Close watch is also kept for any leaks, especially at the beginning of the loading operation.

During loading operations, ballast will be discharged to compensate for the oil being filled into the tanks. This is a very important issue as a lack of ballast compensation can lead to excessive internal stresses being applied to the ship's hull with potential catastrophic consequences. See the Widdy Island disaster case study 19.2.

Case Study 19.2

Widdy Island Disaster 1979 – On 8 January 1979, the oil tanker *Betelgeuse* was discharging its cargo of oil via an offshore jetty at Widdy Island in Southern Ireland. At about 1.00 a.m. a rumbling or cracking noise was heard from the vessel. This was followed a short time later by a huge explosion. A further series of explosions followed which broke the vessel in half. Much of the oil cargo still on board ignited, and this fire continued for a further 12 hours, after which time the vessel sank at her moorings. Fifty people died in the incident.

It was determined that a faulty unloading operation had unbalanced the vessel, causing it to break its back and thereby rupture several of its empty ballast tanks. Vapour from the ruptured tanks then escaped into the vessel and exploded in a fire ball when it found a source of ignition.

The filling of tanks must be monitored, and procedures must be followed which minimize the risks associated with these operations. These include:

▶ The vessel must be securely moored with sufficient mooring scope to ensure it does not range along or away from the berth.

▶ The mooring scope must also take into account tidal rise and fall, river currents and the possible effects of passing ships.

▶ Ensuring hoses are suitable for the product being discharged and the operating pressures they will be subjected to.

▶ Ensuring the connections of pipes and hoses to be used in the transfer operation are secure.

▶ Positioning drip trays beneath all connections and ensuring there is close monitoring of connections during transfer operations.

▶ Deploying fire wires on the vessel to give tugs a means of moving the vessel away from its berth quickly if an emergency arises.

▶ Reducing the risk of static electrical charges occurring on board ship by ensuring all metal objects are bonded to the ship.

▶ Due to the possible differences in electrical potential between the ship and the berth, there is a risk of electrical arcing at the manifold during connection and disconnection of the shore hose or loading arm. To protect against this risk, there should be a means of electrical isolation at the ship/shore interface.

▶ Fire control measures, such as fire-fighting equipment, should be made ready before transfer commences.

▶ Agreement should be reached between ship and shore on a discharge plan which ensures the vessel is not subject to undue internal stresses as the cargo is discharged. (See the Widdy Island disaster case study.)

▶ The control room should monitor flow rates and quantities. This includes alarm systems to indicate when tanks are nearing their filling point. There are also sensors indicating the trim of the marine vessel so that adjustments can be made to the ballast of the vessel as required.

▶ All doors and windows on board the vessel and in buildings at the terminal are to be closed. This is to ensure there is no ingress of flammable vapour which might build up with the potential of causing an explosion.

▶ Adequate venting arrangements should be in place to ensure vapour is dispersed properly and safely. This will include monitoring wind direction and strength. Low wind speed can be an added hazard as the dispersion of vapour in these conditions is minimal and it can build up in dangerous quantities without being apparent. If there is an ignition source nearby, the results can be catastrophic.

▶ Venting arrangements should be made for both the recipient tank and the donor tank. The donor tank will require a volume of air, or more likely inert gas, to replace the volume of product transferred.

▶ Once discharge commences, the vessel must be kept within the operating envelope (limits) of the oil loading arms.

Figure 19.5 Tanker loading at jetty
Source: iStock.

Figure 19.6 Oil loading arms
Source: Dreamstime.

Figure 19.7 Unloading oil from barge
Source: Wise Global Training.

19.3 Control of marine operations, certification of vessels, inspection and approvals

19.3.1 Certification of vessels

The role of the International Maritime Organization (IMO) is to develop international conventions that set out the minimum acceptable standards for the maritime industry. These standards include such things as the construction of the vessel, the safety equipment it must carry, and the training and certification of crew.

The exact interpretation of these conventions into law is determined by each country which has declared its willingness to undertake the registration of vessels. These countries are then known, in a maritime sense, as 'flag states'.

All vessels are required to be registered under a flag state when they are first built, which means the owners of the vessel agree to comply with maritime regulations of the flag state.

During the initial registration process, the vessel is inspected by the flag state inspection team and, provided it meets the required standards, it will be issued with a number of certificates, each relating to various aspects of the vessel.

These certificates are renewed at various intervals on satisfactorily passing a re-inspection. Some certificates are renewable annually and some are renewable every 5 years with an intermediate inspection being conducted after 2½ years.

Owners can, and do, change the flag state of their vessels, and when this happens the vessel is re-inspected under the standards set by the regulations of the new flag state.

19.4 Roles and responsibilities of marine co-ordinators, masters and crew

19.4.1 Roles and responsibilities of masters and crew

On any seagoing vessel, the person responsible for the safety of the vessel and all those aboard is the master. The master has the absolute right – and duty – to make the final decision on matters affecting the vessel and those on board.

The size and structure of the crew will depend on a number of factors, including the size of the vessel, the type of cargo, the requirements of the owner, etc. All ships are issued with a minimum safe manning certificate. However, this is purely for getting a vessel

Table 19.1 Roles, responsibilities and typical numbers of ship's crew

Rank	Responsibility	Reports to	Numbers on board
Master	Overall command	Head office	1
Chief officer/mate	Maintenance of deck. Cargo ops. Bridge watch at sea.	Master	1
Second mate	Navigator. Deck watch in port. Bridge watch at sea.	Master. Cooperates with mate.	1
Third mate	Maintains life-saving equipment and fire-fighting equipment. Deck watch in port. Bridge watch at sea.	Master. Cooperates with mate.	1
Bosun/CPO	Deck crew supervisor	Mate	1
Able bodied seaman/SG1	Skilled deck worker. Rigger. Helmsman.	Bosun	4
Deckhand	Deck worker in training	Bosun	2
Chief engineer	Overall technical maintenance of vessel	Master	1
Second engineer	Engine room maintenance	Chief engineer	1
Third engineer	Power generators	Chief engineer. Cooperates with second engineer.	1
Fourth engineer	Anything to do with fuel	Chief engineer. Cooperates with second engineer.	1
Fitter	Technical maintenance supervisor	Second engineer	1
Pumpman	Technical deck maintenance	Mate and chief engineer	1
Motorman	General engine room maintenance	Fitter	3
Chief steward	In charge of catering and housekeeping	Master	1
Cook		Chief steward	2
Stewards	Housekeeping and waiting on tables	Chief steward	2
Cargo engineer	On Liquefied Natural Gas (LNG) carriers, refrigeration, etc. for cargo	Chief engineer	1

Source: Adapted from www.ilo.org/dyn/normlex/en/f?p=1000:53:0::NO:53:P53_FILE_ID:3130434; and www.maritime-transport.net/mtso/downloads/Public_Information/MTCP_report_safe_manning_level_study.pdf

from A to B and does not take into account such things as managing the cargo, maintenance, etc.

A typical crew list for a seagoing tanker is set out in Table 19.1 above.

19.4.2 Roles and responsibilities of marine co-ordinator

A marine co-ordinator can be based on board a Floating Production, Storage and Offloading (FPSO) unit or at a terminal.

The role involves being responsible for the co-ordination, testing and maintenance of all marine systems and equipment. He/she is also responsible for ensuring all marine activities, procedures and guidance are in compliance with current legislation, codes and standards.

The marine co-ordinator will also lead on cargo transfer operations. These include:

▶ The mooring of the vessels
▶ The connecting and disconnecting of cargo hoses
▶ The unmooring of the vessels
▶ Co-ordinating the use of the support vessel

He/she will also work with the crew of the ship (through the master) to ensure that the transfer of cargo and all other operations are carried out in a safe and efficient manner.

If the marine co-ordinator is based on board an FPSO unit, he/she might also oversee the day to day running of this type of vessel.

In summary, the marine co-ordinator is responsible for:

▶ Liaising between the ship's crew and the asset
▶ Planning, directing and co-ordinating the safe and efficient transfer of product between Liquefied Petroleum Gas (LPG)/oil tankers and the terminal or Floating Production, Storage and Offloading (FPSO) unit
▶ Ensuring the safety of personnel and facilities
▶ Applying timely and appropriate emergency measures when required to do so

19.5 Personnel transfers and boarding arrangements

There are three primary methods of boarding a vessel. These are by:

▶ Gangway
▶ Accommodation ladder
▶ Pilot ladder

We will now have a look at each of these boarding arrangements in a little more detail.

19.5.1 Gangway boarding

Figure 19.8 Gangway boarding
Source: iStock.

19.5.2 Accommodation ladder boarding

Figure 19.9 Accommodation ladder boarding
Source: iStock.

Boarding a vessel by gangway is the main and safest method of boarding a vessel. The gangway should have basic safety features such as handrails, non-slip treads and a suspended cargo net beneath the gangway. If the vessel is a tanker, the Emergency Shutdown (ESD) facility is also positioned adjacent to the gangway.

The gangway will always be manned by a watchman who will be in radio contact with the officer of the watch. The watchman's duties include implementing security procedures as well as keeping and updating a Personnel On Board (POB) list. This is so that if an emergency situation arises, there is accurate knowledge of who needs to be accounted for.

At the shore side of the gangway there will also be a fire plan so that, in case of emergency, shore responders can access a plan of the ship.

An accommodation ladder is an access ladder which is a permanent feature of the ship and is connected to the ship's side. Its elevation can be adjusted according to the requirements of those wishing to board or leave the vessel. This could be at quay level or at sea level. If the ladder is used alongside a quay, a safety net should be rigged below the ladder. Adjustments to the elevation of the ladder are made by an on-board davit.

As with the gangway, when accommodation ladders are used they should be manned at all times.

19.5.3 Pilot ladder boarding

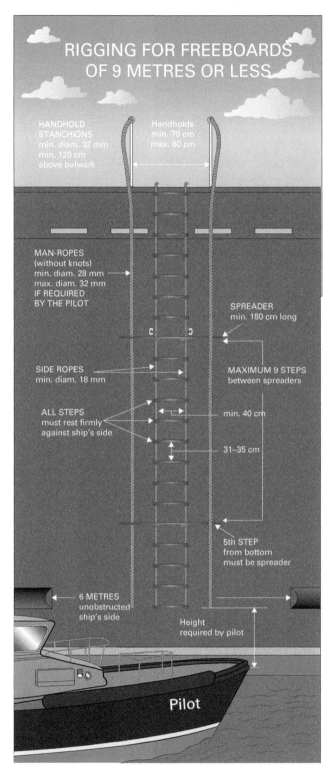

Figure 19.10 Pilot boarding up to 9 metres
Source: International Maritime Pilots Association.

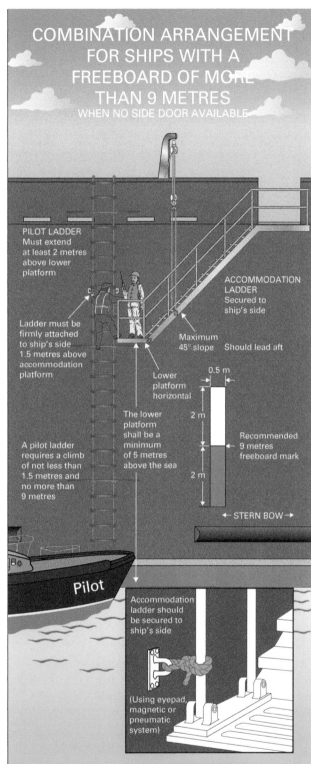

Figure 19.11 Pilot boarding over 9 metres
Source: International Maritime Pilots Association.

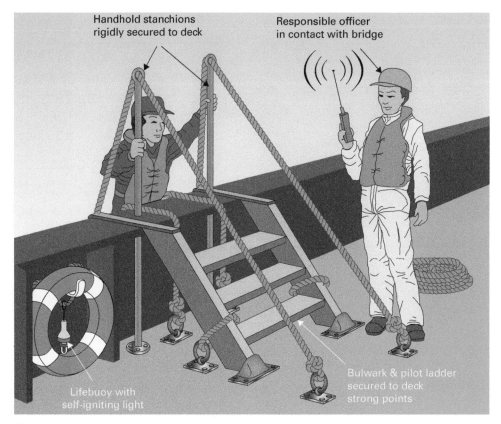

Figure 19.12 Pilot boarding deck arrangements
Source: International Maritime Pilots Association.

As its name suggests, a pilot ladder is for use by pilots to board and leave a ship. It should not be used by any other personnel for transferring between vessels, as it requires training and experience to undertake the transfer safely. If personnel need to board or leave a vessel from an attendant vessel, this should be done via the accommodation ladder.

The main hazard with this pilot ladder transfer is judging the swell of the sea whilst stepping from the ladder onto the other vessel. It also requires an experienced coxswain to ensure the pilot boat does not foul the ladder and cause undue strain which might lead to the pilot ladder being torn from its deck fixing points.

There are strict rules for the construction and use of these ladders. Figures 19.11, 19.12 and 19.13 are taken from a leaflet issued by the International Maritime Pilots' Association.

19.5.4 Other types of personnel transfer

There are two other types of personnel transfer which are occasionally used.

The first is the basket transfer, sometimes known as Billy Pugh. This transfer method relies on the ability of the crane operator as well as the determination of the personnel to hold onto the netting.

Sometimes nicknamed 'Billy Pugh' after its inventor, Captain Billy Pugh, this method of transfer is not internationally accepted, but is still used in some

Figure 19.13 Billy Pugh transfer system
Source: Billy Pugh Co.

parts of the world. Basically, the basket is hoisted and lowered by a crane. The personnel being hoisted stand on the basket's perimeter, lean in towards the net and hold tight before being lifted.

The main risk with this type of transfer is of falling off, which can be controlled by holding tight and using some means of harnessing personnel to the device.

The second type of basket transfer is known as the 'Frog'. Unlike the Billy Pugh, the people being transferred climb inside the frog and get strapped into it. It still requires a very experienced crane operator to ensure safe transfer of personnel. Figure 19.14 shows personnel being transferred by the Frog.

Figure 19.14 Frog transfer system
Source: Reflex Marine.

19.6 Personal protective equipment suitability

The following section describes good practice for selecting protective clothing and other equipment for use in a marine environment.

Personal Protective Equipment (PPE) – specific requirements for marine use

▶ Lifejackets should be used when working over the side or on deck in heavy weather.
▶ Wet weather and/or cold weather clothing should be capable of providing adequate protection. Temperatures as low as −25°C can be experienced and, when coupled with a wind chill factor, can present extreme conditions.
▶ Orinasal masks fitted with suitable filter cartridges should be used on liquefied petroleum gas/liquefied natural gas vessels.
▶ Emergency Life Support Apparatus (ELSA) should be installed at strategic places to aid escape from leaking gas on LPG/LNG carriers.

Personal Protective Equipment (PPE) – training

All personnel should be given adequate training in the use of PPE. Training should cover:

▶ How to properly put on PPE
▶ How to remove contaminated PPE and how to minimize contamination before removal
▶ How to dispose of contaminated PPE safely
▶ The necessity of washing hands after taking off PPE
▶ What to do with damaged PPE
▶ Proper one time use of disposable PPE
▶ Using the correct filters for each different type of gas
▶ Keeping training records

19.7 Diver operations

19.7.1 Diving project plan

Any diving operations will be undertaken by a diving contractor. Consequently, the diving contractor accepts full responsibility for carrying out a safe and efficient dive operation.

Once the extent of the work that requires diver operations to complete has been established and a risk assessment has been carried out covering the work, a diving project plan can be formulated.

The project plan will show how and to what extent the work will be split up into separate dive operations, it will show how the hazards identified in the risk assessment will be controlled, and it will include emergency and contingency plans.

A diving operation is the portion of a diving project identified in the diving project plan which can be supervised safely by one person. The diving contractor will appoint a supervisor to supervise each of these diving operations.

The supervisor has a duty to direct the diving operation safely. If a supervisor does not agree with the size or complexity of the diving project allocated to him/her to supervise, the supervisor has a duty to raise the matter with the diving contractor.

19.7.2 Diving contractor's responsibilities

The diving contractor's general responsibilities are to ensure that:

▶ The diving project is properly and safely managed.
▶ Risk assessments have been carried out.
▶ The place from which the diving is to be carried out is suitable and safe.
▶ A suitable diving project plan is prepared which includes emergency and contingency plans.
▶ The supervisor and dive team are fully briefed on the project and are aware of the contents of the diving project plan.
▶ There are sufficient personnel in the dive team to enable the diving project to be carried out safely.
▶ The personnel are qualified and competent.
▶ Supervisors are appointed in writing and the extent of their control fully documented.
▶ A suitable mobilization and familiarization programme is completed by all the members of the dive team. Other personnel involved in the diving project, for example ship's crew, may also need to complete the programme.
▶ Adequate arrangements exist for first aid and medical treatment.
▶ Suitable and sufficient plant is provided and that it is correctly certified and maintained.
▶ The divers are medically fit to dive.

19

▶ Diving project records are kept containing the required details of the diving project.
▶ There is a clear reporting and responsibility structure laid down in writing.

19.7.3 Diving project – further control measures

Over and above the responsibilities of the diving contractor, there are other basic control measures which should be applied to all diving operations. These include:

▶ Any dive vessel must remain on station throughout the dive operation.
▶ Other subsea work may need to be suspended whilst divers are at work.
▶ All work being conducted above the divers (e.g. construction) should be suspended during the dive operation.

When a dive support vessel is working in close proximity to an installation, a copy of the diving project plan must be passed to the platform. The dive operations are usually covered using the platform's permit-to-work system, written platform procedures and method statements.

Figure 19.15 Dive support vessel
Source: Dreamstime.

REVISION QUESTIONS FOR ELEMENT 5

1 **Explain** what the purpose of an exclusion zone around an offshore installation is and how it can be controlled.
2 The process of loading/unloading vessels with hydrocarbon product is a hazardous operation. **Outline SIX** examples of control measures which could be taken in order to reduce the risks involved in the process.

You can compare your answers with our guidance at the end of the book under the section entitled 'Revision and assessment guide'.

Sub-element 5.2:
Land transport

This chapter covers the following learning outcome

Identify the main hazards of, and suitable controls for, land transport in the oil and gas industries.

20.1   Road tankers

20.1.1   Introduction

Oil and gas have become essential elements in just about every aspect of the way we live and the products we produce. They fuel our transport systems, provide heat and energy to our homes and businesses, and they provide the building blocks for many of the materials and chemicals we use in everyday life.

Consequently, in order to feed this insatiable appetite for oil and gas, vast amounts have to be transported from one location to another at various stages of their transformation to a finished product.

The type of transportation used can vary, and this will depend to a great extent on where the material is coming from and where it is destined to go. Global transportation of hydrocarbons is usually by seagoing tankers, although pipelines are used for some well-established routes. Locally, transportation is by road and rail, and it's these two types of transportation we shall be looking at in this chapter.

20.1.2   Road transportation of hazardous materials

As these products are volatile, and potentially very hazardous, transporting them on public roads needs to be regulated so that safe working practices and procedures, which will help safeguard the public, are universally imposed.

The first step in the safe carriage of hazardous material is to be able to:

▶ Identify exactly what it is that is being carried
▶ Identify what the precise hazards are that this product presents
▶ Identify the best means of controlling those hazards

Conveying this information instantly to those with a vested interest, especially the emergency services, is essential. It also needs to be conveyed in a standard format that is universally recognized by those interested parties.

The UN classification and labelling system for the transportation of dangerous goods, which is now universally accepted, is the means by which this information is made available on all vehicles carrying dangerous goods.

The system uses a standard signboard fixed to the vehicle in a designated position or positions. On this signboard are a number of internationally accepted series of codes and symbols to show what is being transported, the hazards it represents and the correct preventative actions to be taken when required.

Figure 20.1 shows an example of the type of signboard used on vehicles. In this case, the hazard warning sign shows a skull and crossbones, which indicates the product being carried is a toxic substance. The number 6 also indicates it is toxic or there is a risk of infection.

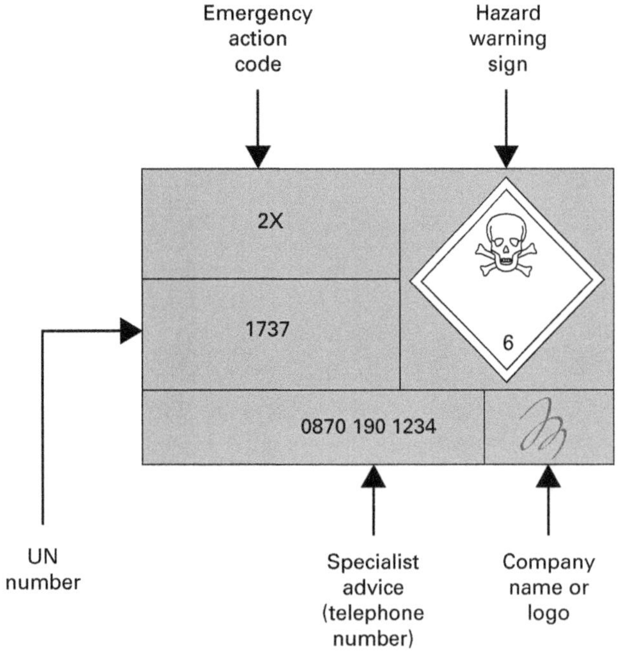

Emergency action code | Hazard warning sign

2X

1737

6

0870 190 1234

UN number | Specialist advice (telephone number) | Company name or logo

Figure 20.1 Hazard signboard panel
Source: Wise Global Training.

The UN classification number 1737 shows the substance being carried in this case is benzyl bromide.

The emergency action code 2X shows the following. The number 2 indicates that the emergency services should use a 'fog or fine spray' when dealing with this substance. The letter X indicates that 'full liquid-tight personal protective equipment' should be worn by the emergency service personnel.

Additional to these information boxes is the telephone number, which will give access to specialist advice if needed.

With regards to transporting hazardous materials, there are nine types of substance classed as hazardous. These are:

1 Explosive substances
2 Flammable gas
3 Flammable liquid
4 Flammable solid
5 Oxidizing agents
6 Toxic substances
7 Radioactive substances
8 Corrosive substances
9 Miscellaneous substances

Figures 20.2 and 20.3 show the hazard warning panel in situ. This panel will be positioned on both sides of the tanker and, as in this case, on the rear of the vehicle as well.

Figure 20.2 Hazard signboard on side of tanker
Source: Wise Global Training.

20.1.3 Protection of plant and equipment against being struck by vehicles

Vehicles travelling at any speed carry the potential to cause significant damage to any object they strike. When that object, or indeed the vehicle itself, contains a hazardous or flammable liquid or gas, the results can be catastrophic.

Consequently, it is essential that all possible measures are taken to reduce this hazard as far as possible.

Figure 20.3 Hazard signboard on rear of tanker
Source: Wise Global Training.

20.1.4 General principles for avoiding plant being struck by vehicles

▷ Traffic routes should be wide enough for the safe movement of the largest vehicle permitted to use them (including visiting vehicles).

▷ Traffic routes should have enough height clearance for the tallest vehicle permitted to use them (including visiting vehicles).

▷ Potentially dangerous obstructions, such as overhead electric cables, or pipes containing hazardous chemicals, need to be protected using goalposts, height gauge posts or barriers.

▷ Traffic routes should be planned to give the safest routes between calling places.

▷ Routes should avoid passing close to such things as unprotected fuel or chemical tanks or pipelines.

▷ All potentially vulnerable plant should be protected from errant vehicles by collision barriers.

▷ All routes should be well lit.

▷ Any hazardous sections of the route, such as sharp bends or adverse cambers, should be clearly signed.

▷ A person should be appointed to be responsible for, and oversee, site traffic movements on site.

▷ Drivers should be trained and authorized to drive vehicles on site.

▷ Visiting drivers should be briefed about site traffic movement rules.

▷ Reversing of vehicles should be avoided or controlled.

20.1.5 Driver training for transportation of dangerous goods

The main objectives of driver training, in relation to carrying dangerous goods, are to ensure drivers:

▷ Are aware of the hazards arising when they are driving a vehicle which is carrying dangerous goods

▷ Know what steps to take in order to reduce the likelihood of an incident taking place

▷ Know what necessary measures they need to take when driving a vehicle which is carrying dangerous goods with regard to ensuring their own safety, that of the public and of the environment; also, if an incident does occur, how to limit the effects of that incident

▷ Have practical experience, attained during the course of training, of what actions they will need to take in the case of an incident occurring

The regulations in some countries require drivers of road tankers to hold a Vocational Training Certificate (VCT) which certifies that the holder has attended appropriate training courses and has passed an examination on the requirements to be met when carrying dangerous goods by road.

The basic course content includes:

▷ General requirements covering the carriage of dangerous goods

▷ The main hazards associated with the carriage of dangerous goods

▷ Preventative and safety measures regarding various hazards related to the carriage of dangerous goods

▷ Actions to be taken in the case of an accident

▷ Marking, labelling, placarding and hazard warning panels

▷ What not to do as a driver of a vehicle carrying dangerous goods

▷ Reasons for, and operation of, technical equipment on vehicles carrying dangerous goods

▷ Precautions to be taken when loading and unloading dangerous goods from vehicles

▷ Restrictions and instructions on driving vehicles carrying dangerous goods through tunnels

Beyond the basic course, tanker drivers will be required to partake in a specialist course which will include:

▷ Behaviour of loaded tankers, including load movement, whilst being driven

▷ Knowledge of different filling and discharge arrangements for tankers

▷ Specific requirements for tankers, including knowledge of certificates of approval for tankers, approval marking, hazard warning panels, etc.

20.1.6 Loading and discharging arrangements

Figure 20.4 Loading arrangements for road tankers
Source: Dreamstime.

Whenever hydrocarbons are transferred from one location to another, as when tankers are loaded or unloaded, there is an increased risk of fire or explosion from a number of sources. Knowing what these sources are, and implementing pre-emptive safety and control measures, helps to keep the risk to a minimum.

The following control measures should be implemented when the loading and unloading of hydrocarbon materials takes place.

- The area designated for loading and unloading should be situated away from general traffic routes. It should also be situated on level ground.
- There should be sufficient space to allow the largest planned-for vehicles to easily manoeuvre into and out of the loading/unloading area.
- Loading/unloading areas should be adequately lit when in use.
- A system should be implemented that ensures a vehicle cannot be driven away from the loading/unloading point before authorized to do so. An example of this control would be to have the driver hand over the ignition key to the transfer supervisor on arrival, who would only returning the key once he/she was satisfied the loading/unloading operation had finished and it was safe for the vehicle to depart.
- When the loading/unloading operation has been completed, drivers must ensure all tank openings, including valves and caps, are closed before starting their journey.
- No tank should be overfilled, as most tanks require room for expansion of the liquid. Information on what is the prescribed level of filling may be affixed to the tank or the vehicle's chassis. Most modern road tankers use a device known as a 'Scully' to protect against overfilling when loading. This device uses a series of probes inside the compartments of the road tanker to detect when the correct amount of product has been loaded and bring the filling operation to an end.
- When filling tanks, the pressure in the tank must be monitored to ensure it does not exceed its maximum working pressure. Tanks should be fitted with pressure relief valves.
- When discharging tanks, the pressure in the tank must be monitored to ensure a vacuum is not created. Tanks should be fitted with vacuum breaker valves.
- All external vents should be fitted with flame arrestors.
- The rate of filling or discharge must be limited. This is to reduce the risk of static electricity build-up, which can be caused by splashing and liquid being circulated unduly.
- Suitable drip trays should be placed beneath hose connection points when loading/unloading operations are being set up or are being ended. This is because these points have the potential to spill product when connection is made or broken.

- Where product with a flash point of 60°C or less is being loaded/unloaded, a bonding wire should be connected between the vehicle and an earthing point before loading/unloading commences.
- A no smoking policy should be established and maintained on site.
- There should be two opposing emergency exits from the loading/unloading area.
- Exits should be clearly marked and open outwards.
- Vapours which are displaced during the transfer operation should be returned to the donor tank via a vapour-tight connection line.
- The vapour return line should have a different connection fitting compared to that of the product transfer hose. This is to ensure there can be no misconnection.
- The vapour return line should be connected before the product transfer hose is connected.
- There should be a device on the vehicle which locks the brakes in the 'on' position when the vapour recovery line is connected.
- A competent person shall be given the responsibility to monitor all the hose connections during loading/unloading operations.
- Any uncontrolled release of vapour should be recorded in the vehicle log book and reported to the authorities.
- There should be a pre-formulated spillage plan ready to deal with any spillages, and a spillage kit at the ready. This kit should include bunding.

20.2   Traffic management

20.2.1   Vehicle movements on site

There is always a need for vehicles to be properly managed and supervised whilst they are on site. Vehicle movements can be managed on site by implementing the following controls:

- A person should be appointed to be responsible for and oversee site traffic movements on site.
- All routes should be well lit.
- Any hazardous sections of the route, such as sharp bends or adverse cambers, should be clearly signed.
- Drivers should be trained and authorized to drive vehicles on site.
- Visiting drivers should be briefed about site traffic movement rules.
- Reversing of vehicles should be avoided or controlled.
- A contingency plan, and the resources to implement it, must be in place in case of adverse weather conditions such as snow and ice.
- Periodic surveys and safety tours should be conducted to ensure traffic rules are being complied with.

20.2.2 Vehicle movements off site

It is equally as important to manage traffic while it's en route by applying control measures which will reduce the chances of an accident or incident occurring. These include the following.

Prior to leaving the site, check that:

▷ All hoses are secure.
▷ There are no leaks from the hoses.
▷ Blanking caps are fitted.
▷ The load is not leaking.
▷ The load is not overheating.
▷ For liquefied petroleum gas tanks, the pressure is within prescribed limits.
▷ The brakes and tyres on the vehicle are OK.
▷ All documents are in order and available.
▷ The markings are correct and in place.

When planning the route, employ the following controls:

▷ Avoid built up areas.
▷ Avoid roads with low bridges.
▷ Obey any restrictions relating to the transport of dangerous goods.

During the journey, check that:

▷ The load is still secure.
▷ There are no leaks.
▷ There is no overheating on the vehicle, i.e. from tyres, bearings, brakes, etc., that could lead to a fire.
▷ Any control temperature is not exceeded.
▷ Markings are still in place, clean and visible.

Employ driving techniques to:

▷ Maintain concentration.
▷ Anticipate situations wherever possible.
▷ Plan ahead.
▷ Drive defensively.
▷ Keep to the speed limit and drive at a safe speed, especially at roundabouts, motorway slip roads and site entrances.
▷ Always allow a margin of safety.

Load behaviour

Drivers must adjust their driving techniques according to the load on board. They must be aware that product will surge backwards under acceleration, will surge forward when braking and will surge from side to side when cornering. Also, any sudden steering movement may well cause instability.

20.3 Rail

The international carriage of dangerous goods by rail within Europe is governed by Annex I of the Convention Concerning International Carriage by Rail. This is known by the letters RID. However, the principles of these regulations form a sound basis for the transport of hydrocarbon products by rail anywhere in the world.

Figure 20.5 Overturned tanker
Source: Dreamstime.

In order to transport hydrocarbon products by rail, a number of duties must be complied with. These include the following:

▷ Any dangerous goods being transported by rail must be clearly marked and labelled with their name, description and UN number, just as with road transport.
▷ Only carriers with the appropriate resources and experience should be engaged to carry hydrocarbon products.
▷ All temporary storage of rail traffic carrying hydrocarbon products should be secure.
▷ Rail staff engaged in the transportation of hydrocarbon products should undergo awareness training programmes.
▷ There should be an emergency plan in place where hydrocarbon products are involved.

When transporting hydrocarbon products by rail, all the safety systems which apply to the safe transportation by road, also apply to rail.

Rail loading/unloading operations present hazards that are very similar to those experienced by road transportation methods. However, there are several additional hazards that are unique to rail. These are:

▷ The movement of trains through entrances and exits of rail sidings should be supervised by a competent person.
▷ Sidings should be isolated from any main line.
▷ During loading/unloading operations, warning signs should be displayed on the train if open to access. These will include:
  ▷ Red flag during the day
  ▷ Red light at night
  ▷ A warning sign that the rail cars are connected.
▷ Loading/unloading operations must be monitored throughout the operation.
▷ Rail cars which have been disconnected from the locomotive must be prevented from moving.
▷ There should be a 15-metre exclusion zone around any loading/unloading point. This exclusion zone will prohibit any potential source of ignition.

Figure 20.6 Loading arrangements for rail tankers
Source: Dreamstime.

▷ Where tools have to be used, these should be non-sparking tools.

▷ Prior to loading/unloading operations commencing, a system of vapour control should be established.

▷ The closure of all foot valves, lids and the removal of hoses, etc. must be overseen by a competent person.

▷ If rail tankers are fitted with product heaters, these must *not* be used if flammable vapours are present during loading and discharge operations.

▷ When work involves working on top of rail cars, working at height regulations should be adhered to.

▷ Weather conditions (the potential for lightning) should be taken into consideration prior to the commencement of loading/unloading.

REVISION QUESTIONS FOR ELEMENT 5 CONTINUED

3 **Outline** what the different sections of a hazard signboard are, as used on road tankers, and what they signify or are for.

4 The process of loading/unloading road tankers with hydrocarbon product is a hazardous operation. **Give SIX** examples of control measures which could be taken in order to reduce the risks involved in the process.

You can compare your answers with our guidance at the end of the book under the section entitled 'Revision and assessment guide'.

Sources of reference for Element 5

Health and Safety Executive (n.d.) 'Ship/Platform Collision Risk in the UKCS – The Regulator's Perspective'. HSE. Available at: www.oilandgasuk. co.uk/downloadabledocs/734/george

Health and Safety Executive (1998) *Commercial Diving Projects Offshore: Diving at Work Regulations 1997– Approved Code of Practice (Legislation).* London: HSE Books. ISBN 978-0-7176-1494-3.

Health and Safety Executive (2002) *Marine Risk Assessment (Offshore Technology Report).* London: HSE Books. ISBN 0-7176-2231-2.

Health and Safety Executive Offshore (n.d.) 'COSSH Essentials – Personal Protective Equipment (PPE) (OCM3) Control Approach 4 Special advice'. HSE. Available at: www.hse.gov.uk/pubns/guidance/ocm3. pdf

International Association of Oil and Gas Producers (2006) *Guideline for Managing Marine Risks Associated with FPSOs: Report No. 377.* London: OGP Publications. Available at: www.ogp.org.uk/ pubs/377.pdf

International Chamber of Shipping, Oil Companies International Marine Forum and IAPH (2006) *ISGOTT: International Safety Guide for Oil Tankers and Terminals, 5th Revised Edition.* London: Witherby & Co. ISBN 978-1-8560-9291-3.

International Maritime Organization (1978) International Convention on Standards of Training, Certification and Watchkeeping for Seafarers (STCW). Available at: www.imo.org/About/Conventions/ ListOfConventions/Pages/International-Convention-on-Standards-of-Training,-Certification-and-Watchkeeping-for-Seafarers-(STCW).aspx

International Maritime Organization (2009) 'Guidelines for the Approval of Fixed Dry Chemical Powder Fire-Extinguishing Systems for the Protection of Ships Carrying Liquefied Gases in Bulk'. Available at: www.imo.org/blast/blastDataHelper.asp?data_ id=25958&filename=1315.pdf

International Maritime Pilots' Association and International Chamber of Shipping (2012) 'Shipping Industry Guidance on Pilot Transfer Arrangements Ensuring Compliance with SOLAS'. 2nd edn. London: Marisec Publications. Available at: www.impahq.org/admin/resources/ pilottransferarrangementsbrochure.pdf

Oil Companies International Marine Forum (OCIMF) (n.d.) 'International Marine Forum Guidelines'. Available at: www.ocimf.com/library/ information-papers

United Nations (1985) 'Annexe 1 Regulations Concerning the International Carriage of Dangerous Goods by Rail (RID)'. UN Treaty Series, vol. 1397, 1-23353, page 171. Available at: http://treaties. un.org/doc/publication/UNTS/Volume%201397/ v1397.pdf

United Nations (2010a) 'Annex B Provisions Concerning Transport Equipment and Transport Operations Part 8: Requirements for Vehicle Crews, Equipment, Operation and Documentation'. Available at: www.unece.org/fileadmin/DAM/trans/danger/publi/adr/adr2011/English/Part8.pdf

United Nations (2010b) 'Requirements for Vehicle Crews, Equipment, Operation and Documentation'. Available at: www.unece.org/fileadmin/DAM/trans/danger/publi/adr/adr2011/English/Part8.pdf

United Nations (2011) 'Globally Harmonized System of Classification and Labelling of Chemicals (GHS)'. 4th revised edn. Available at: www.unece.org/fileadmin/DAM/trans/danger/publi/ghs/ghs_rev04/English/ST-SG-AC10-30-Rev4e.pdf

Walker, S. (n.d.) *Mumbai High North Accident: Presentation to Marine Safety Forum*. HSE Offshore Division. Available at: www.marinesafetyforum.org/upload-files//notices/amm-05.06-mumbai-high-north.pdf

20

Revision and assessment guide

Although this book can be used exclusively as your source of education on the subject of oil and gas operational safety, there are distinct advantages in also undertaking a tutor-supported course to enhance your understanding of the subject matter. The reason for this is that the practical experience of the tutor to be able to explain and discuss with you how the issues covered in this book are applied in real life will inevitably enhance your understanding of the subject and potentially improve your expected examination result. Also, the tutor-supported course material is able to go into issues in much greater depth.

Wise Global Training is able to offer both e-learning and classroom based courses throughout the world, and we would welcome any enquiry regarding delivery of these modes of learning. Our contact details are:

Website: www.wiseglobaltraining.com
E-mail: info@wiseglobaltraining.com
Telephone: 44 (0)1482 211989
Office address: Louis Pearlman Centre,
Goulton Street, Hull, HU3 4DL, UK

When you have completed your study and are ready to take your examination, you will sit a two-hour examination consisting of one paper made up of eleven questions. There will be one main question which will take about 30 minutes to answer, and ten other questions which will take another 90 minutes to answer.

The pass rate is 45 per cent, but if you fail to achieve this mark you will be 'referred', which means you can re-sit the examination as long as it is within five years of the original sitting. There are no limits on how many times you can re-sit the examination.

Revision guide

Once you have read through all of the material, and completed the course if you are undertaking this as additional support, you should review the content to ensure you fully understand the issues covered. During your revision, make notes of the important points so

that you have a summary of the whole subject and can use it as an aide-memoire when preparing for the examination.

We also recommend you purchase a number of past examination papers which are available directly from NEBOSH. NEBOSH also has examiners' reports which can be useful in understanding how examiners view the way in which candidates generally answered the questions. These reports not only provide an excellent guide on the expected answers to questions but also indicate areas of student misunderstanding.

When NEBOSH sets the examination question papers, it uses what are called 'command words' which dictate the way they expect candidates to answer the questions. Failure to answer the questions in the way the command words indicate can cost marks because they will not have been answered appropriately. Understanding the command words in a question is the key to success in answering it. The command word indicates the nature of the answer and the skills being assessed.

The most frequently used command words include:

▷ **Identify** – This is asking for an answer which selects and names a subject. For example **IDENTIFY** three types of non-destructive testing of welds.
▷ **Give** – This is asking for an answer without an explanation. For example **GIVE** an example of . . . ; **GIVE** the meaning of . . .
▷ **Outline** – This is asking for an answer which gives the principle features or different parts of a subject or issue. An exhaustive description is not required. What is sought is a brief summary of the major aspects of whatever is stated in the question.
▷ **Describe** – This is asking for a detailed written account of the distinctive features of a subject. The account should be factual without any attempt to explain. A further definition of **DESCRIBE** as a command word is a picture in word form.
▷ **Explain** – This is asking for the reasoning behind, or an account of, a subject. The command word is testing the candidate's ability to know or understand why or how something happens.

Wise Global Training is accredited to deliver and offer a range of interactive e-learning courses with full tutor support as well as classroom based courses. These include:

NEBOSH International Technical Certificate in Oil and Gas Operational Safety

NEBOSH International General Certificate in Occupational Health and Safety

NEBOSH General Certificate in Occupational Health and Safety

IOSH Managing Safely

IOSH Working Safely

We will gladly respond to any enquiry on how we can meet your training needs. Our contact details are as given above.

Revision questions and suggested outline answers for Element 1

Question 1

Explain what the main purposes of an accident/incident investigation are.

Answer 1

*The command word in this question is **explain**. This requires an answer which gives a clear account of, or reasons for, a subject or issue. Your answer should expand on the following information:*

There are two main reasons why an accident/incident should be investigated. The first reason is to determine the cause of the accident/incident. The second reason is to use the information forthcoming from the investigation to take corrective action in order to avoid it happening again.

Question 2

(a) **Identify** the **FOUR** steps in an accident/incident investigation.

(b) When gathering information following an accident/incident, **identify SIX** sources of information.

Answer 2a

*The command word in this question is **identify**. This requires an answer which selects and names a subject or issue. Your answer should include the following suggested answers:*

Step 1 Gather the information.

Step 2 Analyse the information.

Step 3 Identify the required risk control measures that will reduce the risk.

Step 4 Formulate an action plan and implement it.

Answer 2b

*The command word in this question is **identify**. This requires an answer which selects and names a subject or issue. Your answer should include SIX of the following suggested answers:*

- Victim statements
- Witness statements
- Plans and diagrams
- CCTV coverage
- Process drawings, sketches, measurements, photographs
- Check sheets, permit-to-work records, method statements
- Details of the environmental conditions at the time
- Written instructions, procedures and risk assessments which should have been in operation and followed
- Previous accident records
- Information from health and safety meetings
- Technical information/guidance/toolbox talk sheets
- Manufacturers' instructions
- Risk assessments
- Training records
- Logs

Question 3

(a) **Explain** what a 'near miss' is.

(b) **Explain** why near misses should be investigated.

Answer 3a

*The command word in this question is **explain**. This requires an answer which gives a clear account of, or reasons for, a subject or issue. Your answer should expand on the following information:*

A near miss is any unplanned incident, accident or emergency which did not result in an injury but which could have done in slightly different circumstances.

Answer 3b

*The command word in this question is **explain**. This requires an answer which gives a clear account of, or reasons for, a subject or issue. Your answer should expand on the following information:*

A near miss should be investigated because any recurrence may result in a more serious outcome. Consequently, an investigation should be conducted to determine its cause so that changes can be made to prevent it happening again.

Question 4

Explain the difference between 'vapour pressure' and 'vapour density'.

Answer 4

*The command word in this question is **explain**. This requires an answer which gives a clear account of, or reasons for, a subject or issue. Your answer should expand on the following information:*

Vapour pressure is the process of evaporation where the energy within those molecules at the surface of a liquid is sufficient for those molecules to escape in the form of a vapour. Vapour density, on the other hand, is the measurement of how dense a vapour is in comparison with air.

Question 5

Explain why, when considering the hazards associated with a vapour or gas, the density of the vapour or gas is important.

Answer 5

*The command word in this question is **explain**. This requires an answer which gives a clear account of, or reasons for, a subject or issue. Your answer should expand on the following information:*

Knowing the density of a gas/vapour will determine whether the gas/vapour will rise or fall if it escapes. This will help in determining where any detectors and sources of ventilation should be placed so that any build-up of hazardous gas/vapour can be detected quickly and dealt with appropriately.

Question 6

(a) Identify FOUR hazards associated with refrigerants.
(b) Give TWO control measures relating to refrigerants.

Answer 6a

*The command word in this question is **identify**. This requires an answer which selects and names a subject or issue. Your answer should include FOUR of the following suggested answers:*

▶ Injury from components or material ejected by the high pressure escape
▶ Frostbite injury to skin or eyes where contact with refrigerant is made
▶ Asphyxiation
▶ Possible explosion or fire if the refrigerant is flammable
▶ When certain refrigerant gases burn they can produce other toxic gases
▶ Liquid refrigerants have a very high expansion rate when changing from a liquid to a gas causing overpressure
▶ Refrigerant gases are heavier than air and will slump if the gases are accidently released resulting in potential pockets of gas in voids, drains, ducts, etc.

Answer 6b

*The command word in this question is **give**. This requires an answer without explanation. Your answer should include TWO of the following suggested answers:*

▶ Have pre-arranged procedures in place to deal with any unexpected release of refrigerant.
▶ Ensure personnel do not have to work in confined spaces where there is a risk that refrigerants may be released. This is because of the very real risk of asphyxiation.
▶ Provide ventilation equipment to deal with any potentially high concentrations of refrigerant.
▶ Ensure that procedures are in place that ensure anyone who is exposed to refrigerant gas is immediately moved out of the affected area to a place where they can breathe fresh air and be given oxygen as necessary. They will also need to be medically examined.

Question 7

Identify the **FIVE** steps involved in undertaking a risk assessment.

Answer 7

*The command word in this question is **identify**. This requires an answer which selects and names a subject or issue. Your answer should expand on the following information:*

Step 1 Identify the hazards.
Step 2 Decide who might be harmed and how.
Step 3 Evaluate the risks and decide on precautions.
Step 4 Record the findings and implement them.
Step 5 Review the assessment on a regular basis and update if necessary.

Question 8

Explain the concept of ALARP.

Answer 8

*The command word in this question is **explain**. This requires an answer which gives a clear account of, or reasons for, a subject or issue. Your answer should expand on the following information:*

As low as reasonably practicable (ALARP) relates to reducing risk to a level regarded as low as reasonably possible. This is where an employer should adopt appropriate safety measures unless the cost (in terms of money, time or trouble) is grossly disproportionate to the risk reduction brought about by that particular safety measure or measures. Once all such measures have been adopted, the risks are said to be 'as low as reasonably practicable'.

Question 9

(a) **Explain** what a HAZOP is.
(b) **Identify** who might be included in a HAZOP team and **give** the reason why they should be included.

Answer 9a

*The command word in this question is **explain**. This requires an answer which gives a clear account of, or reasons for, a subject or issue. Your answer should expand on the following information:*

A Hazard and Operability Study (HAZOP) is a tool which is used to systematically examine every part of a process or operation in order to find out how deviations from the normally intended operation of a process can happen, and if further control measures are required to prevent the hazards, which have evolved from the study, from happening.

Answer 9b

*The command words in this question are **identify** and **give**. **Identify** requires an answer which selects and names a subject or issue. **Give** requires an answer without explanation. Your answer should expand on the following information:*

▶ Chairperson – This should be a person who has not been directly involved with the design of the plant, but who has experience in Hazard and Operability (HAZOP) studies.
▶ Design engineer – This will be the person who has been involved with this particular project and will be available to provide information about details of the design.
▶ Process engineer – This will be the person responsible for developing the process and instrumentation diagrams (P&IDs) as well as the process flow diagram.
▶ Electrical engineer – This will be the person responsible for developing the design of the electrical systems within the plant.
▶ Instrument engineer – This will be the person who was concerned with the design and selection of the control systems for the plant.
▶ Operations manager – This will be the person in charge during the commissioning and operation phases of the plant.

Question 10

There are a number of principles involved in inherent safe and risk based design concepts. **Give FOUR** examples.

Answer 10

*The command word in this question is **give**. This requires an answer without explanation. Your answer*

should include FOUR of the following suggested answers:

▶ Minimizing the amount of hazardous material present at any one time
▶ Substituting hazardous materials with less hazardous materials
▶ Moderating the effect a material or process might have (reduce temperature or pressure)
▶ Simplifying the design by designing out problems rather than adding in features to deal with problems
▶ Designing in tolerance levels to cope with faults or deviations
▶ Limiting the effects of any adverse event
▶ Allowing for human error by designing in failsafe features

Question 11

Explain what a safety case/report is.

Answer 11

*The command word in this question is **explain**. This requires an answer which gives a clear account of, or reasons for, a subject or issue. Your answer should expand on the following information:*

A safety case/report is a document which provides evidence and information in order to present a clear, comprehensive and defensible argument that a system is adequately safe to operate in a particular context.

Question 12

In the case of a safety report, **outline** what should be included in an offsite emergency plan.

Answer 12

*The command word in this question is **outline**. This requires an answer which gives the most important features of a subject or issue. Your answer should expand on the following information:*

▶ Details of the site including its location, roads and access points
▶ A site plan showing key facilities such as control centres, medical centres, main process plants and storage areas
▶ Details of site personnel
▶ Details of offsite areas likely to be affected by a major incident as well as levels of possible harm/damage. This will include types of buildings, population density, sensitive buildings, drainage detail, etc.
▶ Details of dangerous substances on site including types of substances, quantities, hazardous properties, etc.
▶ Details of any relevant technical advice
▶ Details of equipment and resources that are available for fire-fighting purposes

- The function of key posts with duties in an emergency response, their location and how they can be identified
- An outline of the initial actions to be taken in case of an emergency situation, such as warning the public, setting up emergency facilities such as a control room, etc.

Revision questions and suggested outline answers for Element 2

Question 1

Identify EIGHT factors that should be considered when selecting a suitable sub-contractor.

Answer 1

*The command word in this question is **identify**. This requires an answer which selects and names a subject or issue. Your answer should include EIGHT of the following suggested answers*:

- Is the contractor adequately insured?
- Has, or will, the contractor undertake a risk assessment of the proposed contracted work?
- Are the health and safety policies and practices of the contractor adequate?
- Is the contractor's recent health and safety performance reasonable (number of accidents, etc.)?
- Is the contractor's health and safety training and supervision adequate?
- Does the contractor have arrangements in place for consulting with their workforce?
- Does the contractor or their individual employees hold a 'passport' or other type of certification in health and safety training?
- What, if any, enforcement notices have been served on the contractor?
- Can the contractor offer any independent assessment of their competence?
- What references from previous employing companies can the contractor show?
- What relevant qualifications and skills does the contractor have?
- What is the level of competency of the staff doing the job?
- Is there appropriate certification for any equipment that the contractor might intend to use?
- What selection procedure does the contractor have for sub-contractors they might engage with?
- Is the contractor a member of a relevant trade or professional body?
- What is the contractor's financial viability?
- Does the contractor's safety method statement meet expectations?

Question 2

When considering plant layout and the possibility of explosion, **explain** what mitigating factors can be introduced at the design stage.

Answer 2

*The command word in this question is **explain**. This requires an answer which gives a clear account of, or reasons for, a subject or issue. Your answer should expand on the following information*:

- Ensuring separation distances are sufficient in even the worst case so that damage to adjacent plants will not occur
- Providing blast walls, or locating adjacent plant in strong buildings
- Ensuring the walls of vessels are thick and strong enough to withstand a pressure wave from an explosion
- Ensuring any explosion relief vents are directed away from vulnerable areas, e.g. other plants or buildings, or roadways near site boundaries

Question 3

Changes to processes and process operations can potentially introduce hazards, some inadvertently. **Explain** what measures can be introduced to ensure changes in procedures are managed as safely as possible.

Answer 3

*The command word in this question is **explain**. This requires an answer which gives a clear account of, or reasons for, a subject or issue. Your answer should expand on the following information*:

- Include expert personnel to review the proposed changes to ensure that they will not result in any operations exceeding established operating limits.
- Ensure that any proposed changes are subject to a safety review using hazard analysis techniques (e.g. hazard and operability studies) to assess the risks. This should be conducted by a team with relevant expertise which covers all areas of the process.
- Have in place arrangements for the control of relevant documents (e.g. safety case/report, drawings, Piping and Instrument Diagrams (P&IDs), etc.) and ensure that they are kept up to date. Any operating procedures should always be updated to reflect any changes in the plant or process.
- Ensure that any changes in the operating envelope (e.g. temperatures, pressures, flow rates, etc.) are communicated to the operators and documented. The operators should also have appropriate training to ensure competent and safe use of the equipment.

Question 4

Explain the role and purpose of a permit-to-work system.

Answer 4

*The command word in this question is **explain**. This requires an answer which gives a clear account of, or reasons for, a subject or issue. Your answer should expand on the following information:*

A permit-to-work is a detailed document which describes specific work at a specific site at a particular time which is to be carried out by authorized personnel. It also sets out any precautions and control measures which are necessary to complete the work safely.

A permit-to-work system allows for competent people to give consideration to the foreseeable risks of a particular task, so that such risks can be reduced to a level that can be regarded as being As Low As Reasonably Practicable (ALARP).

Question 5

Outline FOUR objectives of a permit-to-work system.

Answer 5

*The command word in this question is **outline**. This requires an answer which gives the most important features of a subject or issue. Your answer should include FOUR of the following suggested answers:*

▶ To ensure that proper authorization of designated work is granted

▶ To ensure those people who are conducting the work know the exact nature of the task including hazards, restrictions, time limitations, etc.

▶ To specify the controls and precautions necessary to undertake the work safely, e.g. isolating machinery

▶ To ensure those in charge of the location are aware the work is being carried out

▶ To provide both a system of continuous control and a record that appropriate precautions have been considered and applied by competent persons

▶ To afford the ability to display, to those who need to know, exactly what work is ongoing

▶ To provide a procedural means of suspending work when this is necessary

▶ To provide an ability to control work which might interact, or conflict, with ongoing operations or other permit-to-work activities

▶ To provide a procedural means of handing over the work when that work covers more than one shift

▶ To provide a procedural means of handing back the area or plant which has been involved in the work

Question 6

Describe the basic procedure for implementing a lock out/tag out action.

Answer 6

*The command word in this question is **describe**. This requires an answer which gives a word picture. Your answer should expand on the following information:*

The person authorized to conduct the lock out or tag out action will identify the source(s) of energy to be controlled and the method of control to be used. A safe working procedure will then be established to implement the isolation followed by the lock out, tag out action.

1 All personnel who may be affected by the isolation caused by the lock out, tag out action should be informed.

2 The process or system should be shut down as normal and confirmation made that all controls are in the off position and that all moving parts have stopped.

3 The process or system should be isolated from its source of energy following the safe working procedure set out by the authorized person at the identification of energy stage.

4 Verification that the isolation is effective needs to be made. This is normally done by making an attempt to try to restart the system or process. If it does not start, the isolation has been effective. Once isolation is verified the start-up controls must be returned to their off or neutral position

5 Once the work is completed, the lock out, tag out device will need to be removed.

6 Informing relevant personnel that the plant or process is back in service.

7 The person who has been undertaking the maintenance work should remain whilst the plant or process is restarted.

Question 7

Outline what the process of shift handover involves.

Answer 7

*The command word in this question is **outline**. This requires an answer which gives the most important features of a subject or issue. Your answer should expand on the following information:*

A shift handover should include:

▶ A period of time where the outgoing team prepares the information it will be conveying to the incoming team

▶ A period of time where both the outgoing and incoming team communicate with each other and exchange all relevant information

▶ A period of time where the incoming team cross-checks the information passed on to it as it takes on the responsibility for ongoing operations

Question 8

Identify FOUR of the principles involved in a good shift handover process.

Answer 8

*The command word in this question is **identify**. This requires an answer which selects and names a subject or issue. Your answer should include FOUR of the following suggested answers:*

Shift handover should:

- ▶ Be treated as high priority
- ▶ Not be rushed but be allowed as much time and resource as is necessary to ensure the accurate communication of information
- ▶ Be conducted using both verbal and written means of communication
- ▶ Be conducted face to face, with both parties taking joint responsibility for the effective communication of necessary information
- ▶ Be conducted in an environment which is conducive to good communication without distractions
- ▶ Involve all shift personnel

Question 9

Explain what risk based maintenance is.

Answer 9

*The command word in this question is **explain**. This requires an answer which gives a clear account of, or reasons for, a subject or issue. Your answer should expand on the following information:*

Risk based maintenance is based on increasing the reliability of equipment by assessing the probability of various failure scenarios and applying an appropriate maintenance schedule in order to pre-empt these failures. In the first instance, the likelihood of equipment failure scenarios is formulated.

Then, out of the many likely failure scenarios, the ones which are most credible (most likely to happen) are subjected to an in-depth analysis, including building a detailed consequence picture of each scenario.

From there, a fault tree analysis is drawn up to determine the probability of failure.

The risk of failure is calculated by taking the probability analysis results and combining them with the consequence analysis results.

Question 10

An operator is required to drain water from an LPG sphere.

(a) **Identify** the main hazard associated with this operation.

(b) **Identify** the control measures that can be applied in order to reduce this hazard.

Answer 10a

*The command word in this question is **identify**. This requires an answer which selects and names a subject or issue. Your answer should expand on the following information:*

Hydrate formation can occur when draining water from pressurized Liquefied Petroleum Gas (LPG) spheres. This can lead to valves freezing, and if they are in an open position, LPG will be allowed to escape, which may lead to an explosion.

Answer 10b

*The command word in this question is **identify**. This requires an answer which selects and names a subject or issue. Your answer should expand on the following information:*

The operation of draining water from LPG spheres should only be carried out by trained and authorized personnel using a sequential valve procedure.

Question 11

Thermal shock is regarded as one of the most significant hazards to be considered when a plant or process is started.

(a) **Explain** what thermal shock is.
(b) **Explain** how the effects of thermal shock can be reduced.

Answer 11a

*The command word in this question is **explain**. This requires an answer which gives a clear account of, or reasons for, a subject or issue. Your answer should expand on the following information:*

Thermal shock is where a material is exposed to a sudden and significant change in temperature. This results in the material expanding at different rates within a limited area, causing a crack or failure.

Answer 11b

*The command word in this question is **explain**. This requires an answer which gives a clear account of, or reasons for, a subject or issue. Your answer should expand on the following information:*

- ▶ The gradual introduction of steam or warm product from a lower temperature base (i.e. not superheated steam at the outset)
- ▶ Thoroughly warming up the systems prior to use
- ▶ Designing in expansion loops into the system, for example bellow pieces. These allow the pipework to expand with thermal change without compromising the integrity of the system.

- Using materials with greater thermal conductivity
- Reducing the coefficient of expansion of the materials
- Increasing the strength of the materials

Question 12

There are around twelve stages in the process of commissioning a process plant or system. **Identify SIX** of those stages.

Answer 12

*The command word in this question is **identify**. This requires an answer which selects and names a subject or issue. Your answer should include SIX of the following suggested answers:*

1   The system configuration is checked (walking the line).
2   The pipework and system integrity is checked.
3   The instrumentation system is checked.
4   All alarms are verified as working.
5   All lines and vessels are flushed and cleaned.
6   All ancillary equipment is inspected and assessed as to its adequacy.
7   All instruments and vessels are calibrated.
8   The start-up protocol is established.
9   The shutdown protocol is established.
10   Commissioning trials are undertaken.
11   The plant is hooked up.
12   The plant is handed over.

Revision questions and suggested outline answers for Element 3

Question 1

Explain what stress corrosion cracking is and under what circumstances it might be encountered.

Answer 1

*The command word in this question is **explain**. This requires an answer which gives a clear account of, or reasons for, a subject or issue. Your answer should expand on the following information:*

Stress corrosion cracking is where a material is subjected to both stress and corrosion. The corrosion has the effect of reducing the threshold of the material at a particular point (i.e. it weakens it), resulting in the stress causing the material to crack at that point. If the material had been subject to stress without corrosion, the crack would not have occurred. If the material had been subject to corrosion without stress, again, the crack would not have occurred. It is when these two factors combine that stress corrosion cracking can occur.

Question 2

Define what a safe working envelope is.

Answer 2

*The command word in this question is **define**. This requires an answer which provides a generally recognized or accepted definition. Your answer should expand on the following information:*

A safe operating envelope is defined as the parameters and conditions a plant must operate within to ensure it is not subjected to excessive stress which might introduce or encourage failure modes.

Question 3

Many fixed and floating roof storage tanks have annular rims.

(a)   Outline what an annular rim is.
(b)   Explain what problems can be associated with annular rims.

Answer 3a

*The command word in this question is **outline**. This requires an answer which gives the most important features of a subject or issue. Your answer should expand on the following information:*

The bottom plate in a storage tank is known as an annular plate. The annular plate usually sits on a foundation of hardcore or a concrete ring wall, and is joined to the walls of the tank. This junction between this bottom plate and the wall of the storage tank is known as the annular rim.

Answer 3b

*The command word in this question is **explain**. This requires an answer which gives a clear account of, or reasons for, a subject or issue. Your answer should expand on the following information:*

The joint where the annular plate and the tank walls meet (the annular rim) is subject to constant stress. This is because the weight of the product within the tank wants to push the walls outwards whilst at the same time push the annular plate downwards. This creates a high level of bending stress. The quality of the foundations will also have a bearing on the downward deflection of the annular plate.

A further complicating factor to this stress is the fact that annular plates are prone to corrosion attacks both on the outer side where the tank shell sits on the annular rim, and on the underside of the annular plate where trapped water may lie undetected. This corrosion, coupled with the prolonged stress, can lead to stress corrosion cracking and failure occurring without warning.

Question 4

Identify THREE types of non-destructive testing of welds.

Answer 4

*The command word in this question is **identify**. This requires an answer which selects and names a subject or issue. Your answer should include THREE of the following suggested answers:*

- Magnetic Particle Inspection (MPI)
- Dye Penetrant Inspection (DPI)
- Ultrasonic flaw detection
- Radiography testing

Question 5

Describe what an emergency shutdown system is and its function.

Answer 5

*The command word in this question is **describe**. This requires an answer which gives a word picture. Your answer should expand on the following information:*

Because the petrochemical industry processes large quantities of hazardous material within a contained environment, it needs to have in place systems which will either prevent loss of containment from happening or mitigate the consequences of such an event if it does happen. These systems are known as Emergency Shutdown (ESD) systems and take the form of various components, each of which is designed to deal with a particular scenario and bring it under control in a safe and effective manner.

Question 6

Identify THREE typical actions an emergency shutdown system might perform if activated.

Answer 6

*The command word in this question is **identify**. This requires an answer which selects and names a subject or issue. Your answer should include THREE of the following suggested answers:*

- Shutdown of part systems and equipment
- Isolate hydrocarbon inventories
- Isolate electrical equipment
- Stop hydrocarbon flow
- Depressurize/blow down
- Activate fire-fighting controls (water deluge, inert gas, foam system, water mist)
- Activate emergency ventilation control
- Close watertight doors and fire doors

Question 7

Explain why and when an emergency shutdown system might be bypassed.

Answer 7

*The command word in this question is **explain**. This requires an answer which gives a clear account of, or reasons for, a subject or issue. Your answer should expand on the following information:*

All emergency shutdown systems and fire and gas systems need to be tested, inspected and/or maintained on a regular basis to ensure they are functioning as required. These testing and/or maintenance procedures involve the temporary bypassing of safety system interlocks which would otherwise activate the emergency shutdown system when testing, inspection or maintenance activities were undertaken.

Question 8

Describe the purpose and function of an interceptor.

Answer 8:

*The command word in this question is **describe**. This requires an answer which gives a word picture. Your answer should expand on the following information:*

Interceptors are a means of collecting contaminated water before it is discharged to a foul drain or surface drain. Typically, interceptors have three separate chambers, with the divisions between chambers extending down to the bottom, and low level pipes connecting the chambers. This is so that when the contaminated water enters the first chamber it can separate (oil will naturally float on top of water) and be extracted. The water is then directed to the second and third chambers via the low level pipe where any residual oil is also allowed to separate and be extracted. Finally, the water from chamber three is channelled into either a foul or surface drain, whichever is appropriate.

Question 9

With regard to fixed and floating roof storage tanks, **explain** what erosion is and where it might be expected to occur.

Answer 9

*The command word in this question is **explain**. This requires an answer which gives a clear account of, or reasons for, a subject or issue. Your answer should expand on the following information:*

Erosion is the process of material being worn away by the constant movement of product flowing over the surface. Areas such as filling and discharge points, which experience large amounts of product flow, are the most vulnerable points.

Question 10

Explain what hazards are associated with fixed roof storage tanks.

Answer 10

*The command word in this question is **explain**. This requires an answer which gives a clear account of, or reasons for, a subject or issue. Your answer should expand on the following information:*

Normally, tanks have a pressure-vacuum relief valve fitted on or near the top of the tank. This allows vapour or gas to escape or, in the case of a reduction in pressure within the tank, allows air to enter the tank. Consequently, the pressure-vacuum relief valve is a critical piece of equipment and the integrity of the tank is dependent upon it working properly.

The weight of any liquid pumped into a tank will naturally apply outward pressure on the walls of the tank it is being stored in. Also, as the liquid builds up inside the tank, the atmosphere in the void needs to be vented out. If this does not happen the pressure will become unsustainable and the tank may rupture.

The contra view is also true in that when a tank is emptied the liquid reduces and the void above the liquid needs to have air vented in so that a vacuum is not created. If the air inlet valve is faulty, and a vacuum is created, this may well cause the tank to collapse.

There are other factors which can cause increased or decreased pressure within the tank. These include:

▶ Storing a volatile product will cause gases to evolve and increase pressure.
▶ Warm weather or direct sun on the tank will warm the product up and make it expand, thus increasing pressure.
▶ Cold weather will cool the product down and cause it to contract, thus decreasing pressure.

Question 11

With regard to floating roof storage tanks,

(a) **Explain** what 'landing the roof' is and what hazards are associated with it.
(b) **Explain** what control measures need to be taken if the roof is landed and why.

Answer 11a

*The command word in this question is **explain**. This requires an answer which gives a clear account of, or reasons for, a subject or issue. Your answer should expand on the following information:*

Landing the roof is where the liquid in the tank falls far enough for the legs on the underside of the floating roof to land on the base of the tank. The void between the liquid and the roof will grow, which will allow a build-up

of vapour to occur. This has the potential to cause a fire and/or an explosion. Another issue with landing the roof is that it causes a particular corrosion mechanism called fretting-related corrosion. This is where repeated contact on the base of the tank by the legs on the underside of the floating roof removes protective layers of rust scale that may have formed, resulting in increased corrosion rates.

Answer 11b

*The command word in this question is **explain**. This requires an answer which gives a clear account of, or reasons for, a subject or issue. Your answer should expand on the following information:*

If the roof is landed, it is important that enough time elapses to allow the space below the roof to become saturated with vapour so that it exceeds its upper flammable limit. This is because when product is pumped into the tank and the roof begins to lift, there is a chance that a spark may be generated between the tank wall and the roof seal. Having the vapour in the ullage saturated to beyond its upper flammable limit will ensure the vapour will not be ignited by any errant spark. The amount of time required to allow the vapour to become saturated may be as much as 24 hours.

Question 12

Explain what a 'bund' is and its purpose and function.

Answer 12

*The command word in this question is **explain**. This requires an answer which gives a clear account of, or reasons for, a subject or issue. Your answer should expand on the following information:*

Where storage tanks hold a liquid product there is always the possibility that the tank will rupture and spill its contents. Consequently, contingencies should be in place to control such a situation. This is where a containment wall, known as a bund, is built around a tank, or group of tanks to contain any spillage that may occur until such time that it can be dealt with. The bund also stops the product spreading too far, seeping into the ground, or seeping into the drainage or water systems. It also reduces (although it does not eliminate) the risk of fire and/or explosion.

Question 13

Define what a BLEVE is and **describe** an example of how a BLEVE might develop.

Answer 13

*The first command word in this question is **define**. This requires an answer which provides a generally recognized or accepted definition. The second command word in this question is **describe**. This* requires an answer which gives a word picture. Your answer should expand on the following information:

A BLEVE is a Boiling Liquid Expanding Vapour Explosion. An example of how a BLEVE might develop is when a vessel is partly filled with pressurized hydrocarbon liquid (e.g. Liquefied Petroleum Gas). It will have a certain amount space above it filled with vapour, and if the vessel is subjected to a fire the pressure in the tank will increase due to the liquid going above its boiling point and turning into a vapour. The pressure relief valve on the tank will allow the overpressure to be vented to atmosphere in the first instance, but this will reduce the amount of liquid in the tank still further, and the potential for the flame to engage with a section of the tank containing vapour and not liquid will increase. If this happens, the tank wall will weaken at this point as the heat transfer to vapour is much less efficient than it is to a liquid. The result is likely to be a sudden and catastrophic failure of the vessel, with a discharge of vapour followed by an explosion when the vapour reaches the flames. This is a Boiling Liquid Expanding Vapour Explosion (BLEVE).

Question 14

Electrostatic charges are a problem in hydrocarbon process systems in that they can cause sparks with the potential to create an explosion.

(a) **Identify THREE** factors which can influence the generation of electrostatic charges.
(b) **Give THREE** measures that can be used to reduce the generation of electrostatic charges.

Answer 14a

The command word in this question is **identify**. This requires an answer which selects and names a subject or issue. Your answer should include THREE of the following suggested answers:

Factors which can influence the generation of static electricity include the following:

- The conductivity of the liquid
- The amount of turbulence in the liquid
- The amount of surface area contact between the liquid and other surfaces
- The velocity of the liquid
- The presence of impurities in the liquid
- The atmospheric conditions. Static build-up is enhanced when the air is dry.

Answer 14b

The command word in this question is **give**. This requires an answer without explanation. Your answer should include THREE of the following suggested answers:

Measures that can be used to reduce the generation of electrostatic charges include:

- Ensure that filling operations do not involve the free-fall of liquids.
- Lower the velocity of the liquid being filled.
- Ensure fill pipes touch the bottom of the container being filled.
- Tanks which have been filled with products that have a low conductivity, i.e. jet fuels and diesels, should be given time to relax before the process continues.
- Tanks which have been filled with product should not have any ullage (vapour space) for a set period of time. Nor should any dipping of the product take place, again for a set period of time.

Question 15

With regard to hazardous areas,

(a) **Explain** why hazardous areas are categorized into different zones.
(b) **Explain** why equipment should be categorized for use in different zones.

Answer 15a

The command word in this question is **explain**. This requires an answer which gives a clear account of, or reasons for, a subject or issue. Your answer should expand on the following information:

Gases and vapours can create explosive atmospheres. Consequently, areas where these potentially hazardous airborne substances present themselves are classed as hazardous areas so that appropriate controls can be implemented. However, the frequency with which these substances present themselves is also a factor in determining the appropriate level of control. For example, if the presence of a flammable vapour only happens once every three months, it would not be sensible to apply the same level of control as to an area where a flammable vapour is present all day, every day. The answer is to apply a classification to areas – called zoning – which places appropriate controls on the type of equipment that can be used in that area and which potentially can create a source of ignition, particularly electrical equipment, which reflects the risk involved.

Answer 15b

The command word in this question is **explain**. This requires an answer which gives a clear account of, or reasons for, a subject or issue. Your answer should expand on the following information:

The issue with electrical equipment is that it normally creates sparks, either as part of the brushes coming in contact within the rotating armature, or when a switch is activated. Either event can ignite any flammable gas present in the atmosphere in the vicinity of the equipment.

Consequently, manufacturers have designed specialized equipment which overcomes, in various ways, the issue of having sparks which are exposed to the local atmosphere. The particular solution which is incorporated into each piece of equipment is signified by a code which is marked on the equipment's product identification label.

Question 16

There are two main types of boiler used to heat water. **Explain** what they are and how both types work.

Answer 16

*The command word in this question is **explain**. This requires an answer which gives a clear account of, or reasons for, a subject or issue. Your answer should expand on the following information:*

A fire tube boiler is one which generates hot gases which then pass through a number of tubes before being expelled out of the flue. These tubes run through a sealed and insulated container of water and the heat from the gases is transferred by thermal conductivity to the water, which then turns to steam. The steam from the boiler then exits through a tube at the top of the container.

Water tube boilers have basically the opposite configuration of fire tubes. In a water tube boiler, a number of tubes run through the furnace part of the boiler. This heats the water inside the tubes, which turns it to steam.

Question 17

In relation to boilers and furnaces, **explain** what flame impingement is and the potential consequences if it is not rectified

Answer 17

*The command word in this question is **explain**. This requires an answer which gives a clear account of, or reasons for, a subject or issue. Your answer should expand on the following information:*

Flame impingement is where the flame produced by the burner, within the boiler or furnace, comes into contact with the surface being heated. This is something which the design of the boiler or furnace does not normally allow. This is because the temperature of the surface being heated should always remain within prescribed parameters and direct flame contact can lead to these parameters being exceeded. If this occurs there tends to be a gradual build-up of carbon on the inside of the tube at the point where the flame is in contact. This layer of carbon acts as an insulator between the water/product inside the tube and the surface where the flame is impinging on it, compounding the effect the flame is having on the heated surface of the tube.

If this process is allowed to continue, it can lead to the tube eventually becoming blocked, resulting in the potential rupture of the tube.

Question 18

Explain what total dissolved solids are and how they can be controlled.

Answer 18

*The command word in this question is **explain**. This requires an answer which gives a clear account of, or reasons for, a subject or issue. Your answer should expand on the following information:*

Total dissolved solids are substances, such as minerals, salts and metals, which are held in a suspended form within water. If these solids are of a sufficient concentration within the water used in a boiler system, they can attach themselves to the inside of boilers and, over time, build up to form scale.

The first action in controlling the effects of total dissolved solids is to maintain the solids below a certain limit with the water. This is done by testing the water with a TDS meter or conductivity meter. This measures the conductivity of the water which is an indication of the measure of total dissolved solids within the water. The greater the concentration of total dissolved solids in the water, the greater the likelihood is that the dissolved solids will precipitate out of the water and form scale on the inside of the boiler. If concentration levels are deemed to be too high, it is necessary to drain some of the water from the system, called boiler blow down, in order to remove some of those dissolved solids and keep the total dissolved solids concentration below the level where they will precipitate.

Revision questions and suggested outline answers for Element 4

Question 1

Describe how an infrared gas detector system works

Answer 1

*The command word in this question is **describe**. This requires an answer which gives a word picture. Your answer should expand on the following information:*

Infrared absorption combustible gas detectors work by being able to recognize the specific absorption characteristics that hydrocarbon molecules have to infrared light. They send an infrared beam between a fixed transmission device and a fixed receiving device. The more hydrocarbon molecules that are present in a given space being monitored by these devices, the higher the absorption of infrared radiation which is detected by the receiving device.

Question 2

Explain what passive fire protection is.

Answer 2

*The command word in this question is **explain**. This requires an answer which gives a clear account of, or reasons for, a subject or issue. Your answer should expand on the following information:*

The term 'passive fire protection' covers the materials, products and design measures built into a building or structure in order that any fire which may start in the building or structure is restricted in its growth and spread. This is achieved by controlling the flammability of the structure, including walls, ceilings, floors, doors, etc., as well as by protecting structural steel members from severe heat which might compromise their integrity. Finally, the building or structure is designed in a way that divides it into separate fire containing compartments. This is so that any fire which may start will be restricted, or contained within a limited area.

Question 3

With regard to passive fire protection, **explain** the difference between a wall coded as H120 and a wall coded as A30.

Answer 3

*The command word in this question is **explain**. This requires an answer which gives a clear account of, or reasons for, a subject or issue. Your answer should expand on the following information:*

Walls and divisions which have been manufactured as passive fire protection are coded in relation to the type of fire they are designed to withstand and its duration. The code is made up of a letter (either H or A) followed by a number (0, 30, 60 or 120). The letter indicates what type of fire the wall or division has been made to withstand. 'H' type fire walls are manufactured to a standard that will withstand hydrocarbon fires. Whereas 'A' type fire walls are manufactured to a standard that will only withstand cellulosic fires. The number which follows the letter indicates the length of time the wall or division has been designed to hold back the fire. For example, a wall with a 'H120' coding is designed to withstand a hydrocarbon fire for 120 minutes, whilst a wall with an 'A30' coding is designed to withstand a cellulosic fire for 30 minutes.

Question 4

Foam has a number of differing effects on a fire. **Give THREE** of these effects and **explain** how each of them affects a fire.

Answer 4

*The first command word in this question is **give**. This requires an answer without explanation. The second command word in this question is **explain**. This requires an answer which gives a clear account of, or reasons for, a subject or issue. Your answer should include THREE of the following suggested answers:*

▶ **A separating effect** – The foam cover separates the combustion zone from the surrounding atmosphere and, as such, prevents oxygen feeding the fire.
▶ **A covering effect** – The foam cover stops evaporation of flammable vapours from the burning material. This means flammable vapours, which would normally be encouraged by the heat of the fire and which feed the fire, are stopped at source.
▶ **A cooling effect** – When foam is applied to a fire which is fed from a liquid source of fuel, water is gradually discharged by the foam which adds a cooling effect to the flame.
▶ **A suppression effect** – High- or medium-expansion foam used to flood areas will suppress the release of flammable vapours/gases and so prevent them combining with air which is necessary for the combustion process to take place.
▶ **An insulating effect** – Foam has a low thermal conductivity level. Consequently, when foam covers any flammable liquid which is not burning, it tends to insulate it from thermal radiation and ignition.
▶ **A film formation effect (AFFF – aqueous film forming foam)** – Foam which is used on non-polar hydrocarbons (the majority of hydrocarbons) produces a thin aqueous film which helps the foam flow, as well as assisting in the extinguishing process and inhibiting re-ignition.

Question 5

Describe an appropriate fire protection system that could be used on a floating roof storage tank.

Answer 5

*The command word in this question is **describe**. This requires an answer which gives a word picture. Your answer should expand on one of the following systems:*

System (a)

Floating roof storage tanks have a linear heat detection system at the rim seal area of the roof and this will be connected to a number of foam based fire extinguishing modules positioned around the perimeter of the roof. When the linear heat detector registers a fire, as well as raising the alarm it will also activate the fire extinguisher module nearest to the fire. This will then flood the immediate area with foam and extinguish the fire.

System (b)

Floating roof storage tanks have a linear heat detection system at the rim seal area of the roof, and this will be connected to a fire extinguishing system which has a number of foam outlets positioned around the top wall of the tank. These are connected by pipework to a foam mixing tank positioned outside the bunded area. This mixes and pumps the foam through the pipes to the outlets where the foam is discharged to flood the full perimeter of the tank. The foam can be supplemented by the fire service, if required, as the mixing tank will have a foam inlet for the emergency service to add additional foam.

Question 6

Give FOUR factors which can enhance chances of survival and escape from an installation in an emergency situation.

Answer 6

*The command word in this question is **give**. This requires an answer without explanation. Your answer should include FOUR of the following suggested answers:*

- Escape routes should be clearly marked using high visibility signage along their entire route.
- Escape routes should be well lit and include emergency lighting in case power is cut off.
- All escape routes should be protected, where possible, by fire walls or by deluge fire protection systems.
- Escape routes should be congestion-free (allow clear access and egress) and be adequate in size to accommodate all personnel.
- The installation should be compartmentalized (have fire walls between compartments).
- Blow out, or explosion panels, should be strategically positioned within the installation to alleviate any overpressure.
- Escape routes should have heat activated deluge/sprinklers within them.
- Each area of the installation should have more than one escape route.
- Temporary Refuge (TR) buildings should be constructed from material which has a fire rating of at least H120.
- Temporary Refuge (TR) buildings should be under positive atmospheric pressure using an airlock system.
- There should be more than one means of communicating specific instructions to personnel, such as what to do and where to go.
- The number of ways an alarm is conveyed to personnel should not be by siren alone, i.e. a flashing beacon should be used for areas where a siren might not be heard.

- Multiple means of manually descending to sea level should be provided, i.e. knotted rope, sea ladder attached to the platform leg, scramble net, skyscape, etc.
- Appropriate personnel escape equipment should be available either in the accommodation area or on each escape route. This equipment should include Emergency Breathing Systems (EBS), Emergency Life Support Apparatus (ELSA), smoke hoods, torches and flame retardant gloves.

Question 7

Explain what emergency drills are and who should participate in them.

Answer 7

*The command word in this question is **explain**. This requires an answer which gives a clear account of, or reasons for, a subject or issue. Your answer should expand on the following information:*

Drills, in relation to emergency situations, are exercises which are undertaken to evaluate emergency plans and procedures under realistic conditions. They also serve the purpose of training personnel in the practical application of their roles and responsibilities. Consequently, they should be conducted as often as is deemed necessary in order to maintain competency levels. Involving external emergency services and agencies in these exercises is considered essential in order to facilitate an efficient and effective response should the need arise. Offshore, muster drills should be undertaken weekly. However, these drills should be random, unannounced, and undertaken at varying times of the shift in order to prove they are effective under all foreseeable circumstances.

Revision questions and suggested outline answers for Element 5

Question 1

Explain what the purpose of an exclusion zone around an offshore installation is and how it can be controlled.

Answer 1

*The command word in this question is **explain**. This requires an answer which gives a clear account of, or reasons for, a subject or issue. Your answer should expand on the following information:*

The immediate area around an installation is regarded as a major risk area with potentially severe consequences for any incident that happens within it. Consequently, it is designated as an exclusion zone with restrictions on which vessels are allowed to enter the zone. The zone extends for 500 metres around the installation and is constantly radar-monitored, as well as being patrolled by the platform's standby vessel or its emergency response and rescue vessel, either of which is in close communication with the platform's central control room. Any vessel wishing to enter the exclusion zone must seek permission from the central control room and the patrol vessel before doing so. If a vessel tries to enter the zone without permission, it will be warned off by the patrol vessel.

Question 2

The process of loading/unloading vessels with hydrocarbon product is a hazardous operation. **Outline SIX** examples of control measures which could be taken in order to reduce the risks involved in the process.

Answer 2

The command word in this question is **outline**. *This requires an answer which gives the most important features of a subject or issue. Your answer should include SIX of the following suggested answers:*

▶ The vessel must be securely moored with sufficient mooring scope to ensure it does not range along or away from the berth. The mooring scope must also take into account tidal rise and fall, river currents and the possible effects of passing ships.

▶ Ensuring hoses are suitable for the product being discharged and the operating pressures they will be subjected to.

▶ Ensuring the connections of pipes and hoses to be used in the transfer operation are secure.

▶ Positioning drip trays beneath all connections and ensure there is close monitoring of connections during transfer operations.

▶ Deploying fire wires on the vessel to give tug boats a means of moving the vessel away from its berth quickly if an emergency arises.

▶ Reducing the risk of static electrical charges occurring on board ship by ensuring all metal objects are bonded to the ship.

▶ Due to the possible differences in electrical potential between the ship and the berth, there is a risk of electrical arcing at the manifold during connection and disconnection of the shore hose or loading arm. To protect against this risk, there should be a means of electrical isolation at the ship/shore interface.

▶ Fire control measures, such as fire-fighting equipment, should be made ready before transfer commences.

▶ Agreement should be reached between ship and shore on a discharge plan which ensures the vessel is not subject to undue internal stresses as the cargo is discharged.

▶ The control room should monitor flow rates and quantities. This includes alarm systems to indicate when tanks are nearing their filling point. There are

also sensors indicating the trim of the marine vessel so that adjustments can be made to the ballast of the vessel as required.

▶ All doors and windows on board the vessel and in buildings at the terminal should be closed. This is to ensure there is no ingress of flammable vapour which might build up and potentially cause an explosion.

▶ Adequate venting arrangements should be in place to ensure vapour is dispersed properly and safely. This will include monitoring wind direction and strength.

▶ Venting arrangements should be made for both the recipient tank and the donor tank. The donor tank will require a volume of air, or more likely inert gas, to replace the volume of product transferred.

▶ Once discharge commences, the vessel must be kept within the operating envelope (limits) of the oil loading arms.

Question 3

Outline what the different sections of a hazard signboard are, as used on road tankers, and what they signify or are for.

Answer 3

The command word in this question is **outline**. *This requires an answer which gives the most important features of a subject or issue. Your answer should expand on the following information:*

▶ There is a hazard warning sign/symbol which shows the internationally recognized diamond shaped symbol to indicate what hazard the product presents and its classification.

▶ There is the UN classification number which indicates exactly what the product is.

▶ There is the emergency action code indicating what actions and precautions the emergency services should take if they have to deal with an incident involving the product.

▶ There is a telephone number to contact if specialist advice is required.

Question 4

The process of loading/unloading road tankers with hydrocarbon product is a hazardous operation. **Give SIX** examples of control measures which could be taken in order to reduce the risks involved in the process.

Answer 4

The command word in this question is **give**. *This requires an answer without explanation. Your answer should include SIX of the following suggested answers:*

▶ The area designated for loading and unloading should be situated away from general traffic routes. It should also be situated on level ground.

- There should be sufficient space to allow the largest planned-for vehicles to easily manoeuvre in and out of the loading/unloading area.
- Loading/unloading areas should be adequately lit when in use.
- A system should be implemented that ensures a vehicle cannot be driven away from the loading/unloading point before authorized to do so.
- When the loading/unloading operation has been completed, drivers must ensure all tank openings, including valves and caps, are closed before starting their journey.
- No tank should be overfilled as most tanks require room for expansion of the liquid.
- When filling tanks, the pressure in the tank must be monitored to ensure it does not exceed its maximum working pressure. Tanks should be fitted with pressure relief valves.
- When discharging tanks, the pressure in the tank must be monitored to ensure a vacuum is not created. Tanks should be fitted with vacuum breaker valves.
- All external vents should be fitted with flame arrestors.
- The rate of filling or discharge must be limited to reduce static electricity build-up.
- Suitable drip trays should be placed beneath hose connection points when loading/unloading operations are being set up or are being ended.
- Where product with a flash point of 60°C or less is being loaded/unloaded, a bonding wire should be connected between the vehicle and an earthing point before loading/unloading commences.
- A no smoking policy should be established and maintained on site.
- There should be two opposing emergency exits from the loading/unloading area.
- Exits should be clearly marked and open outwards.
- Vapours which are displaced during the transfer operation should be returned to the donor tank via a vapour-tight connection line.
- The vapour return line should have a different connection fitting compared with that of the product transfer hose. This is to ensure there can be no misconnection.
- The vapour return line should be connected before the product transfer hose is connected.
- There should be a device on the vehicle which locks the brakes in the 'on' position when the vapour recovery line is connected.
- A competent person should be given the responsibility to monitor all the hose connections during loading/unloading operations.
- Any uncontrolled release of vapour should be recorded in the vehicle log book and reported to the authorities.
- There should be a pre-formulated spillage plan ready to deal with any spillages, and a spillage kit at the ready.

Index